Health Sciences
C. Ed.

The Bassett Atlas of Human Anatomy

Robert A. Chase, M.D.

Emile Holman Professor of Surgery and Anatomy
Chairman, Division of Human Anatomy

Stanford University School of Medicine

Dissections by David L. Bassett, M.D.

Photographs by William B. Gruber

The Benjamin/Cummings Publishing Company, Inc.
Redwood City, California • Fort Collins, Colorado
Menlo Park, California • Reading, Massachusetts • New York
Don Mills, Ontario • Wokingham, U.K. • Amsterdam • Bonn
Sydney • Singapore • Tokyo • Madrid • San Juan

Sponsoring Editor: Connie Spatz
Developmental Editor: Darcy Lanham
Production Editor: Larry Olsen
Designer: Gary Head
Color Separation: Color Tech
Printing and Binding: Kingsport Press
Cover Illustration: Rebecca Schwartz

Library of Congress Cataloging-in-Publication Data

Chase, Robert A. (Robert Arthur), 1923–
    The Bassett atlas of human anatomy.

    Includes index.
    1. Anatomy, Human—Atlases.   I. Bassett,
David Lee, 1913–   . II. Title.
[DNLM: 1. Anatomy—atlases.   QS 17 C487b]
QM25.C44   1989   611   88-33302
ISBN 0-8053-0118-6

    BCDEFGHIJ-KP-89

The Benjamin/Cummings Publishing Company, Inc.
390 Bridge Parkway Suite 102
Redwood City, California 94065

Preface

Through the eyes of a master teacher and the lens of an accomplished photographer, students can explore—and understand—the marvels of the human body. These extraordinary photographs offer the next best thing to actual observation.

This unprecedented human anatomy atlas preserves the work of the late David L. Bassett, M.D., of Stanford University, who devoted his career to conducting meticulous human dissections. He collaborated with photographer William B. Gruber, and together they recorded over 1,600 human dissection views for use in teaching.

With the guidance of over 80 experienced anatomy instructors, I was able to prune this vast collection to the 86 photographs that will be most valuable to undergraduate students of anatomy and physiology. Because the photos presented here are limited to those that will best serve students' needs, this atlas provides an inexpensive alternative to the comprehensive and costly collections now available. These photographs will serve as a useful reference to graduates, postgraduates, and professionals as well. Clear line drawings elucidate the more detailed photographs; labels and descriptive legends focus attention on important structures.

To accommodate the various approaches to this course, a Systems Approach Grid is provided. This grid will allow you to tailor the atlas to your course organization.

Slide Package

Qualified adopters of *The Bassett Atlas of Human Anatomy* will be eligible to receive a set of 35 mm color slides that correspond to the photographs in the atlas.

*This book is respectfully dedicated to
the late David L. Bassett, M.D.,
and the late William B. Gruber.*

Contents

1   The Central Nervous System

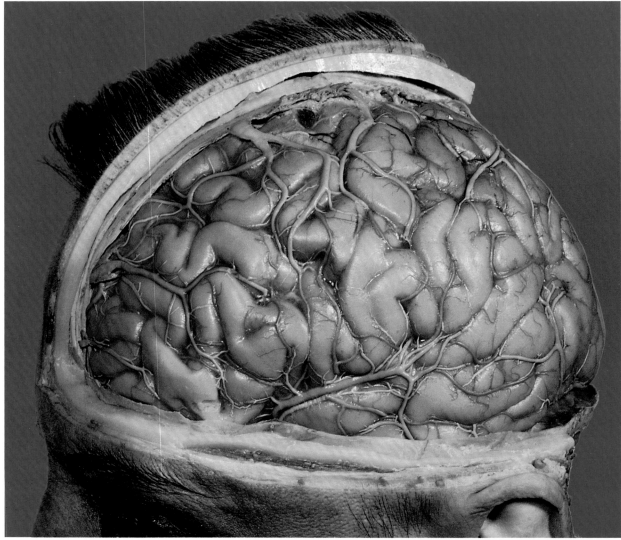

Figure 1.1A   Brain surface and vessels.

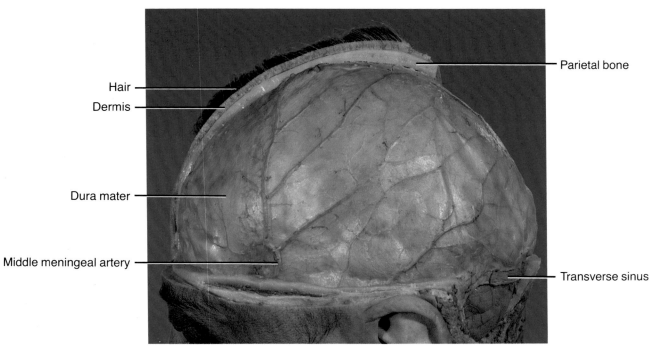

Hair

Dermis

Dura mater

Middle meningeal artery

Parietal bone

Transverse sinus

Figure 1.2   Meninges and vessels.

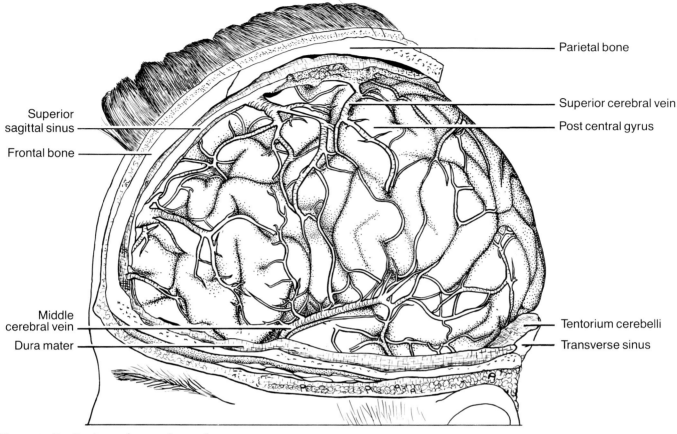

Parietal bone

Superior cerebral vein

Post central gyrus

Superior sagittal sinus

Frontal bone

Middle cerebral vein

Dura mater

Tentorium cerebelli

Transverse sinus

Figure 1.1B   Brain surface and vessels.

Figure 1.1A, B   Brain surface and vessels. The cerebral hemisphere has been exposed by careful removal of the dura to maintain the arachnoid membrane intact. Beneath the arachnoid membrane, one sees the cerebral hemisphere, the cerebral veins, and the cortical arteries. The superior sagittal sinus has been opened along the superior midline. The superior cerebral veins empty into this sinus. The anterior cerebral veins are represented by the middle cerebral vein, which follows the lateral fissure to empty into the cavernous sinus. The tentorium cerebelli is visible on the right, separating the cerebrum from the cerebellum.

Figure 1.2   Meninges and vessels. In this lateral view of the dura mater and meningeal vessels, the calvarium has been removed on the left side. Peeling the internal periosteum (endocranium) from the inner aspect of the skull has preserved the meningeal vessels. Posteriorly, the dura has been removed from the cerebellum, and the transverse sinus (a dural venous sinus) may be seen as it courses along the posterior margin of the tentorium cerebelli. The relationships of the scalp and underlying cranial bones can be seen at the apex.

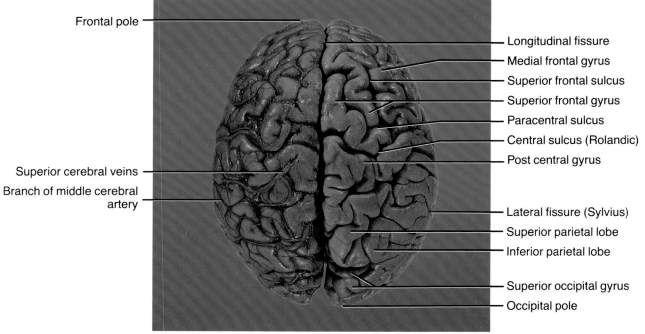

Frontal pole

Longitudinal fissure

Medial frontal gyrus

Superior frontal sulcus

Superior frontal gyrus

Paracentral sulcus

Central sulcus (Rolandic)

Post central gyrus

Superior cerebral veins

Branch of middle cerebral artery

Lateral fissure (Sylvius)

Superior parietal lobe

Inferior parietal lobe

Superior occipital gyrus

Occipital pole

Figure 1.3   Brain surface with and without vessels.

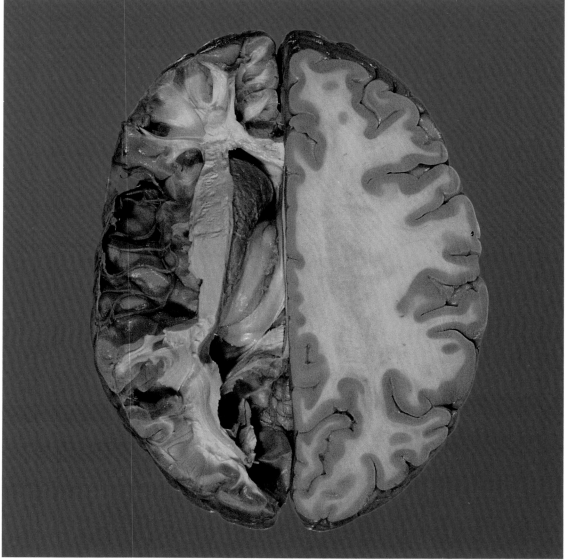

Figure 1.4A   Cross section of the brain.

Figure 1.3 Brain surface with and without vessels.
This general surface view of the brain from its superior aspect shows the surface meninges and vessels removed from the right hemisphere to demonstrate the configuration of the cortical gyri and sulci. On the left, the arachnoid membrane is intact, except near the midline. Branches of the middle cerebral artery ramify over the lateral surface of the left hemisphere, and branches of the posterior cerebral artery may be seen posteriorly. Several branches of the anterior cerebral artery are visible medially and anteriorly. Larger superior cerebral veins and points of entry into the venous lacunae of the dura have been cut off. They are filled with blue latex for better visualization. Major sulci and gyri are identified on the right hemisphere.

Figure 1.4A, B Cross section of the brain. The brain has been sectioned and is viewed superiorly. The right cerebral hemisphere has been sectioned horizontally just above the level of the corpus callosum. Note the differentiation between gray and white matter. Important relationships of surface landmarks to deeper structures may be clearly seen in this view. Note the lateral ventricle and its choroid plexus on the left side.

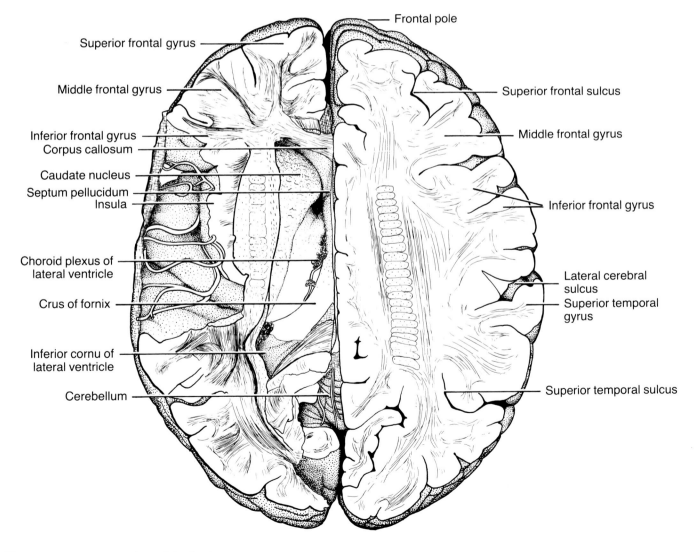

Figure 1.4B Cross section of the brain.

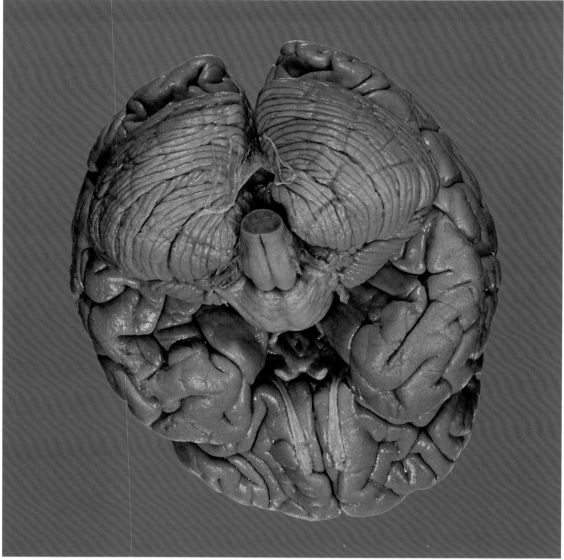

Figure 1.5A   Base of the brain.

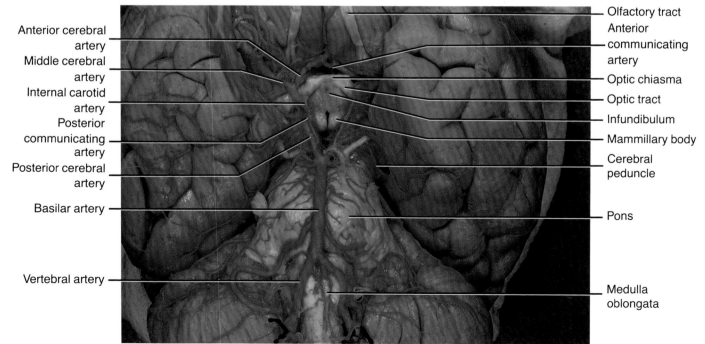

Anterior cerebral artery

Middle cerebral artery

Internal carotid artery

Posterior communicating artery

Posterior cerebral artery

Basilar artery

Vertebral artery

Olfactory tract

Anterior communicating artery

Optic chiasma

Optic tract

Infundibulum

Mammillary body

Cerebral peduncle

Pons

Medulla oblongata

Figure 1.6   Circle of Willis.

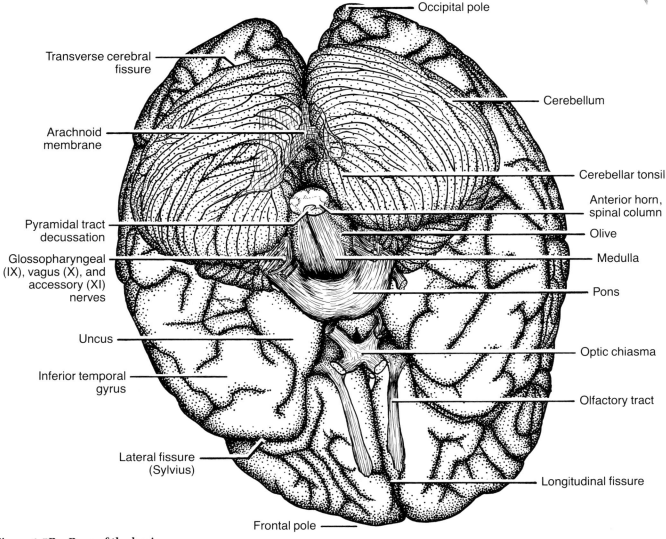

Occipital pole

Transverse cerebral fissure

Cerebellum

Arachnoid membrane

Cerebellar tonsil

Anterior horn, spinal column

Pyramidal tract decussation

Olive

Glossopharyngeal (IX), vagus (X), and accessory (XI) nerves

Medulla

Pons

Uncus

Optic chiasma

Inferior temporal gyrus

Olfactory tract

Lateral fissure (Sylvius)

Longitudinal fissure

Frontal pole

Figure 1.5B   Base of the brain.

Figure 1.5A, B   Base of the brain. In this view, the base of the brain may be seen with the brainstem sectioned through the decussation of pyramids. The specimen is tilted to illustrate the maximum number of structures.

Figure 1.6   Circle of Willis. This view shows the arteries on the basal surface of the brain. The arterial Circle of Willis is formed by branches of the vertebral and internal carotid arteries. The vertebral arteries give rise to the basilar artery at the caudal end of the Circle, and branches of the internal carotid arteries join the major divisions of the basilar artery to complete the Circle. Part of the right temporal lobe has been removed from the field. This allows one to see the middle cerebral artery with some of its branches in the lateral fissure.

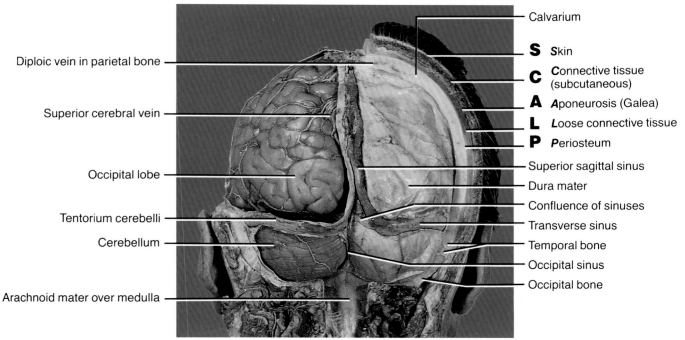

Calvarium

S *S*kin

C *C*onnective tissue (subcutaneous)

A *A*poneurosis (Galea)

L *L*oose connective tissue

P *P*eriosteum

Superior sagittal sinus

Dura mater

Confluence of sinuses

Transverse sinus

Temporal bone

Occipital sinus

Occipital bone

Diploic vein in parietal bone

Superior cerebral vein

Occipital lobe

Tentorium cerebelli

Cerebellum

Arachnoid mater over medulla

Figure 1.7   Posterior view of the brain in situ.

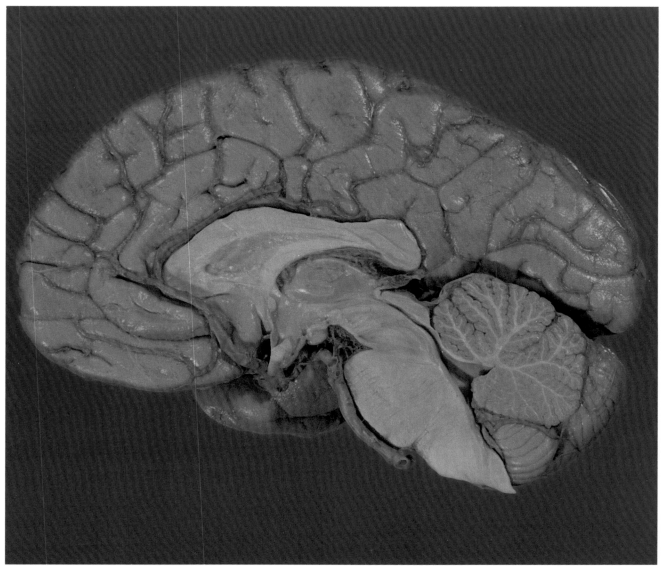

Figure 1.8A   Sagittal section of the brain.

Figure 1.7 Posterior view of the brain in situ. In this posterior view of cranial meninges, the calvarium and layers of the scalp are shown in relation to the dura mater. On the left, the dura has been removed from the field; the cerebral hemisphere and cerebellum are covered by the arachnoid membrane. Note the tentorium cerebelli with the transverse sinus in its posterior margin. The superior sagittal sinus, occipital sinus, and two transverse sinuses meet at the confluence of sinuses. The arachnoid mater confining the cerebrospinal fluid is clearly seen over the medulla. This is the area of the cisterna magna.

Figure 1.8A, B Sagittal section of the brain. The medial surface of the cerebral hemisphere and midline structures of the brainstem and cerebellum are displayed in this medial sagittal section of brain. The third ventricle, cerebral aqueduct, fourth ventricle, and central canal for circulation of cerebrospinal fluid are exposed. The cut surface of the corpus callosum may also be seen. Other divided structures are the optic chiasma, the infundibulum and floor of the third ventricle, the intermediate mass of the thalamus, and the pineal body.

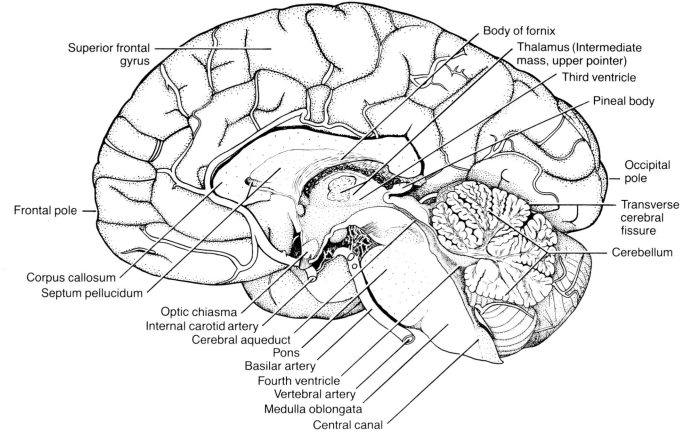

Figure 1.8B Sagittal section of the brain.

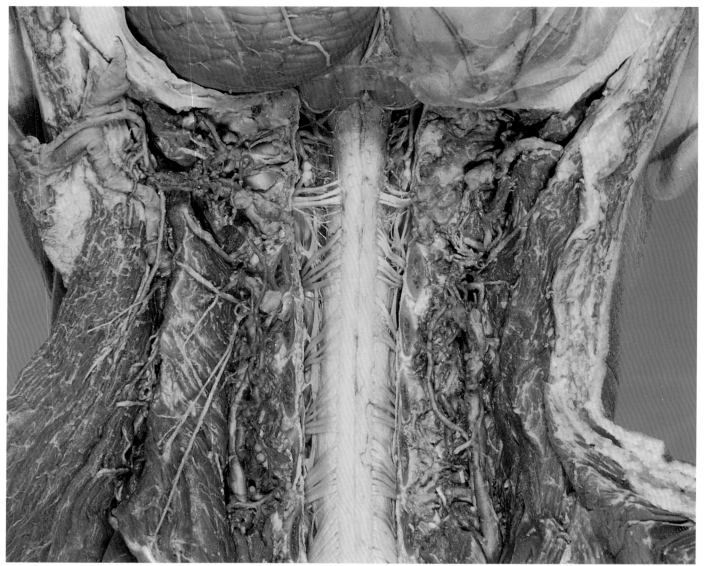

Figure 1.9A Spinal cord origin.

First lumbar vertebral arch
(cut aross)

Iliocostalis muscle

Spinous process of second
lumbar vertebra

Latissimus dorsi muscle

Cauda equina

Longissimus muscle

Conus medullaris

Filum terminale

Figure 1.10 Spinal cord, cauda equina.

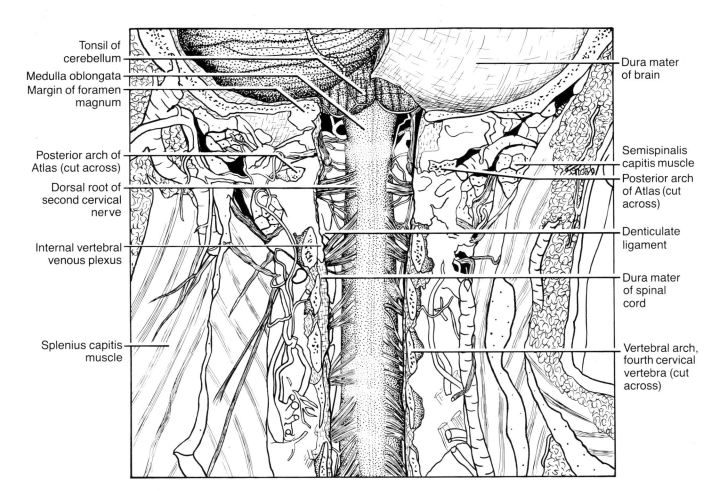

Tonsil of cerebellum

Medulla oblongata

Margin of foramen magnum

Posterior arch of Atlas (cut across)

Dorsal root of second cervical nerve

Internal vertebral venous plexus

Splenius capitis muscle

Dura mater of brain

Semispinalis capitis muscle

Posterior arch of Atlas (cut across)

Denticulate ligament

Dura mater of spinal cord

Vertebral arch, fourth cervical vertebra (cut across)

Figure 1.9B   Spinal cord origin.

Figure 1.9A, B   Spinal cord origin. This view of the cervical spinal cord with the arachnoid membrane removed clearly reveals the cervical posterior roots of the spinal cord. Note the relationship of the tonsils of the cerebellum to the margin of the foramen magnum. Denticulate ligaments of the pia mater support the spinal cord within its meninges. Note the internal vertebral venous plexus around the spinal cord.

Figure 1.10   Spinal cord, cauda equina. In this view of the distal end of the spinal cord, the cauda equina, conus medullaris, and filum terminale are seen in relationship to the first lumbar vertebral arch. The conus medullaris ends at the level of the arch of the second lumbar vertebra.

Figure 1.11A   Spinal cord, detail.

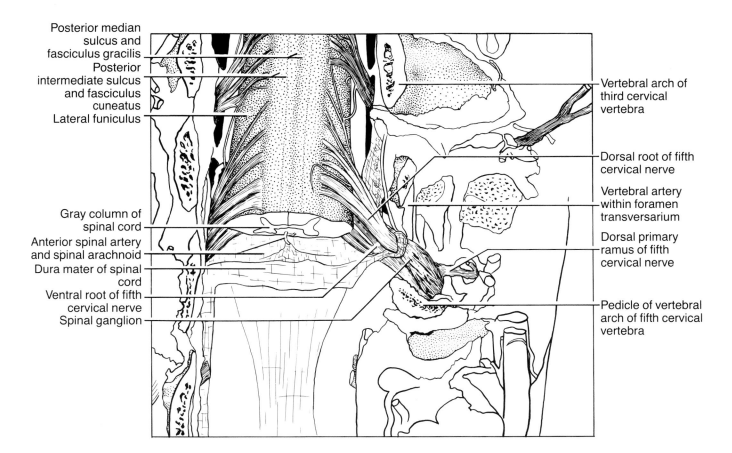

Posterior median sulcus and fasciculus gracilis

Posterior intermediate sulcus and fasciculus cuneatus

Lateral funiculus

Gray column of spinal cord

Anterior spinal artery and spinal arachnoid

Dura mater of spinal cord

Ventral root of fifth cervical nerve

Spinal ganglion

Vertebral arch of third cervical vertebra

Dorsal root of fifth cervical nerve

Vertebral artery within foramen transversarium

Dorsal primary ramus of fifth cervical nerve

Pedicle of vertebral arch of fifth cervical vertebra

Figure 1.11B   Spinal cord, detail.

Figure 1.11A, B   Spinal cord, detail. The spinal cord at the cervical level has been divided between the fifth and sixth cervical segments. The cervical vertebrae have been removed on the right to expose the nerve roots, ganglion, and dorsal primary ramus of the fifth cervical nerve. The gray column of the spinal cord may be seen where the cord is sectioned. The vertebral artery within the foramen transversarium is readily visible.

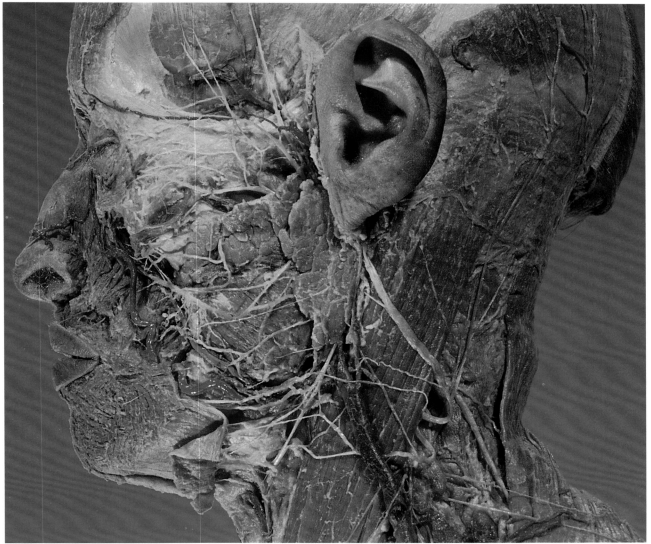

Figure 2.1A   Parotid gland and facial nerve.

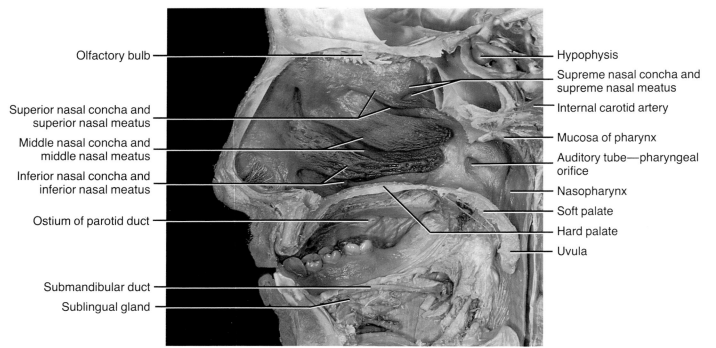

Olfactory bulb

Superior nasal concha and
superior nasal meatus

Middle nasal concha and
middle nasal meatus

Inferior nasal concha and
inferior nasal meatus

Ostium of parotid duct

Submandibular duct

Sublingual gland

Hypophysis

Supreme nasal concha and
supreme nasal meatus

Internal carotid artery

Mucosa of pharynx

Auditory tube—pharyngeal
orifice

Nasopharynx

Soft palate

Hard palate

Uvula

Figure 2.2   Sagittal section, oral and nasal cavities.

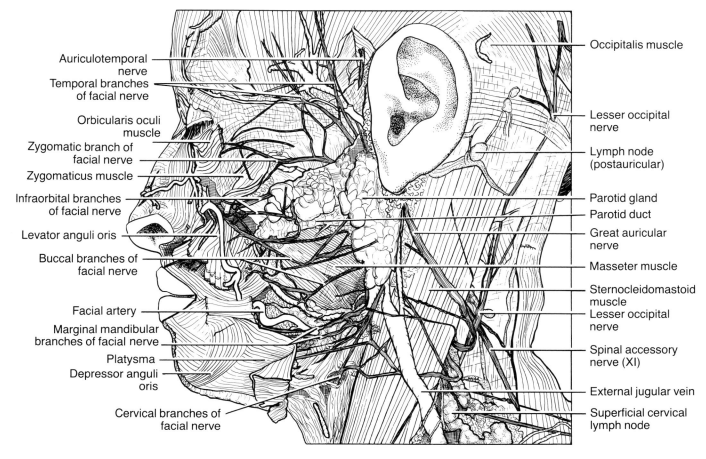

Auriculotemporal nerve

Temporal branches of facial nerve

Orbicularis oculi muscle

Zygomatic branch of facial nerve

Zygomaticus muscle

Infraorbital branches of facial nerve

Levator anguli oris

Buccal branches of facial nerve

Facial artery

Marginal mandibular branches of facial nerve

Platysma

Depressor anguli oris

Cervical branches of facial nerve

Occipitalis muscle

Lesser occipital nerve

Lymph node (postauricular)

Parotid gland

Parotid duct

Great auricular nerve

Masseter muscle

Sternocleidomastoid muscle

Lesser occipital nerve

Spinal accessory nerve (XI)

External jugular vein

Superficial cervical lymph node

Figure 2.1B   Parotid gland and facial nerve.

Figure 2.1A, B   Parotid gland and facial nerve. This dissection of the side of the face and head reveals the parotid gland and its duct over the surface of the masseter muscle. Emerging on its anterior surface from between the deep and superficial lobes are the branches of the facial nerve—temporal, zygomatic, infraorbital, buccal, mandibular, and cervical. This view shows some of the muscles of facial expression innervated by the facial nerve as well as branches of the nerves from the cervical plexus. The spinal accessory nerve (XI) is seen crossing the posterior triangle of the neck.

Figure 2.2   Sagittal section, oral and nasal cavities. This dissection of the nasal fossa, nasopharynx, and palate displays the olfactory bulb and the nasal conchae. Note the auditory tube and its relationship to the hard and soft palates.

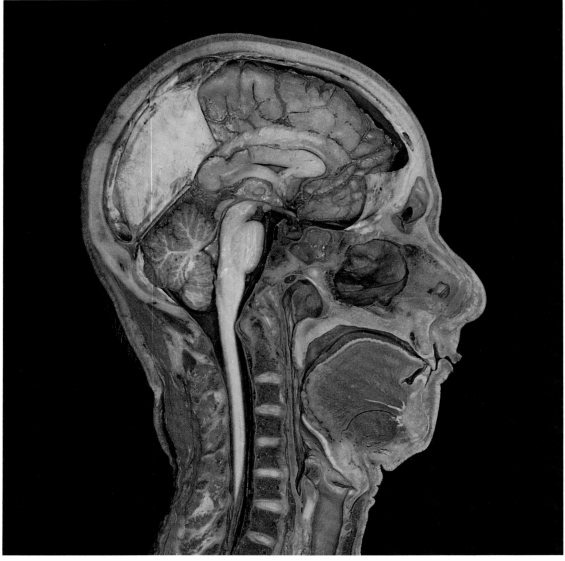

Figure 2.3A Sagittal section, head and neck.

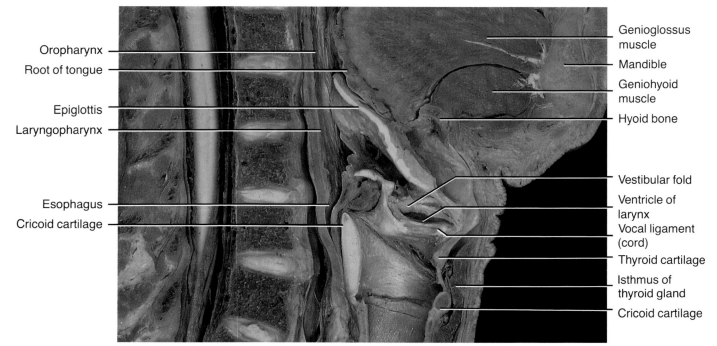

Oropharynx

Root of tongue

Epiglottis

Laryngopharynx

Esophagus

Cricoid cartilage

Genioglossus
muscle

Mandible

Geniohyoid
muscle

Hyoid bone

Vestibular fold

Ventricle of
larynx

Vocal ligament
(cord)

Thyroid cartilage

Isthmus of
thyroid gland

Cricoid cartilage

Figure 2.4 Sagittal section, larynx.

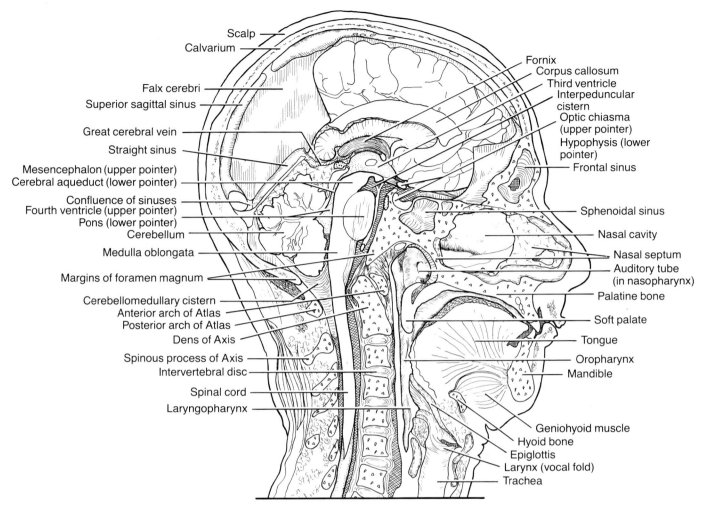

Figure 2.3B   Sagittal section, head and neck.

Figure 2.3A, B   Sagittal section, head and neck. This mid-sagittal section of the head and neck displays elements of the upper nasal respiratory system, including the nasal cavity, nasopharynx, oropharynx, and laryngopharynx. Note the relations of the auditory tube in the nasopharynx to the nasal cavity, palate, and skull base. Midline elements of the central nervous system, some of its venous sinuses, and the elements of the cerebrospinal fluid circulatory system are evident. Note the areas common to the aerodigestive tracts in the oral cavity, oropharynx, and laryngopharynx, where the trachea and esophagus separate. The relationship of the pharynx to the vertebral column, which occupies a near central position in the neck, is clearly shown.

Figure 2.4   Sagittal section, larynx. In this sagittal section of the neck, the mucous membrane of the larynx is removed, except for that which lies in the ventricle and strips on the vocal and ventricular folds. The anatomical position of the epiglottis, to serve its capital role in separation of the digestive and respiratory tracts at the laryngopharynx, is evident. The sectioned hyoid bone, thyroid cartilage, and cricoid cartilage show their relationships, forming an anterior protective wall for the laryngopharynx and larynx.

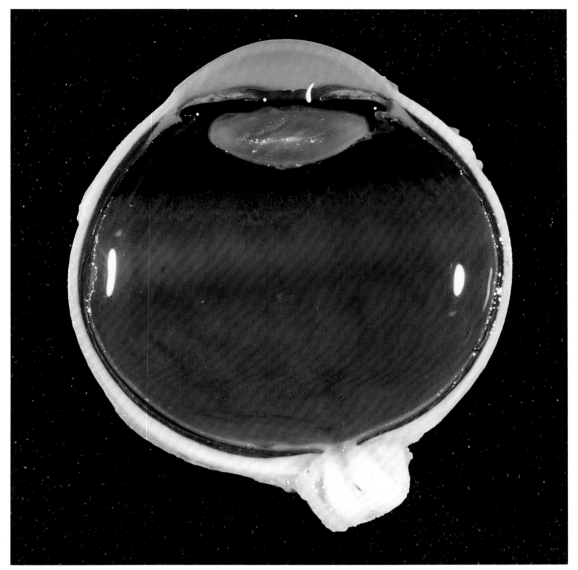

Figure 2.5A   Eye.

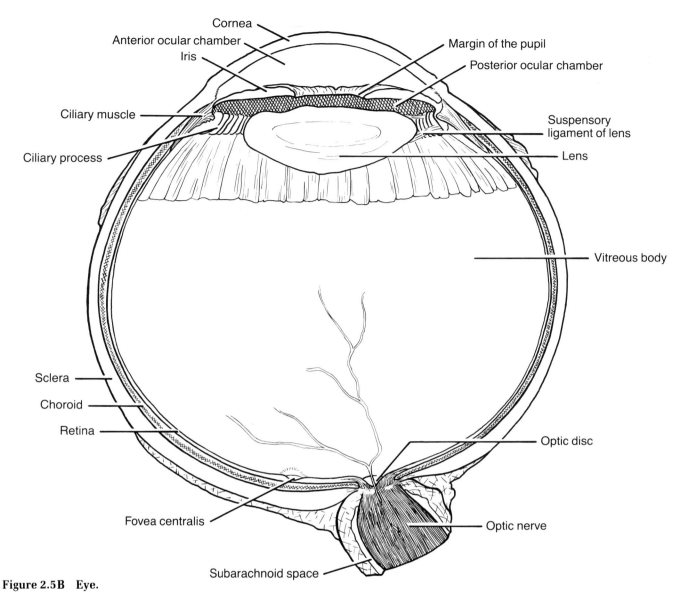

Cornea
Anterior ocular chamber
Iris
Ciliary muscle
Ciliary process
Margin of the pupil
Posterior ocular chamber
Suspensory ligament of lens
Lens
Vitreous body
Sclera
Choroid
Retina
Optic disc
Fovea centralis
Optic nerve
Subarachnoid space

Figure 2.5B   Eye.

Figure 2.5A, B   Eye. This horizontal section through the human eye through the fovea centralis displays the constituents of the globe. The retina is an extension of the optic nerve. The subarachnoid space surrounds the optic nerve all the way to its penetration into the globe.

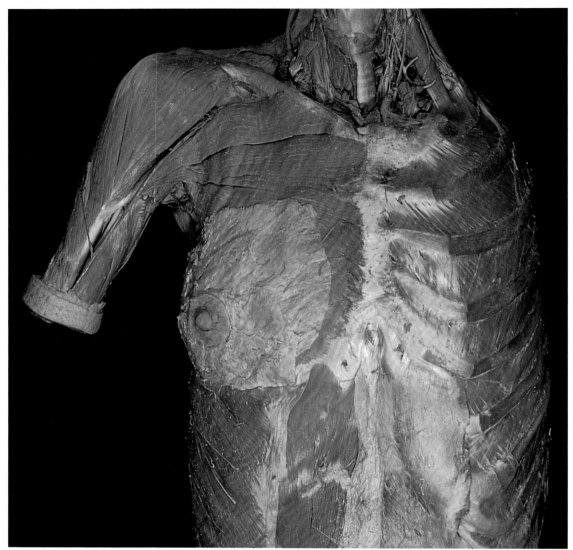

Figure 3.1A   Chest wall and breast.

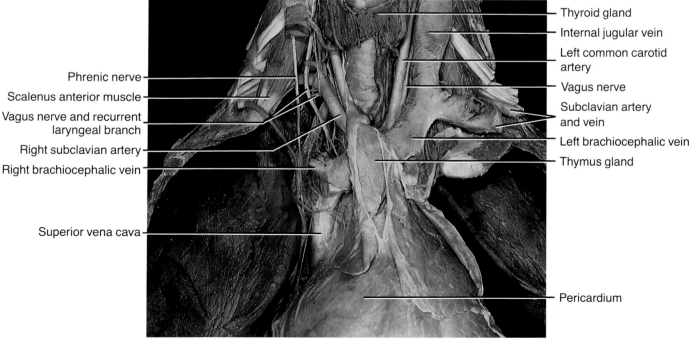

Phrenic nerve

Scalenus anterior muscle

Vagus nerve and recurrent laryngeal branch

Right subclavian artery

Right brachiocephalic vein

Superior vena cava

Thyroid gland

Internal jugular vein

Left common carotid artery

Vagus nerve

Subclavian artery and vein

Left brachiocephalic vein

Thymus gland

Pericardium

Figure 3.2   Anterior mediastinum.

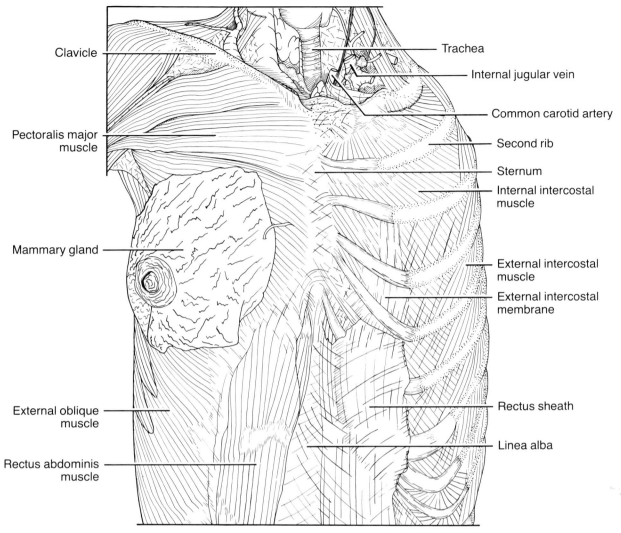

Clavicle

Pectoralis major
muscle

Mammary gland

External oblique
muscle

Rectus abdominis
muscle

Trachea

Internal jugular vein

Common carotid artery

Second rib

Sternum

Internal intercostal
muscle

External intercostal
muscle

External intercostal
membrane

Rectus sheath

Linea alba

Figure 3.1B   Chest wall and breast.

Figure 3.1A, B   Chest wall and breast. Dissection of the breast and anterior thoracic wall displays the relationship of the breast to the pectoral fascia. The pectoralis muscles have been kept in situ on the right. On the left, the clavicle, scapula, and associated muscles have been removed to reveal the thoracic wall. In the first and second intercostal spaces, the external intercostal muscle has been removed to reveal the internal intercostal layer. The external intercostal muscle may be seen in the lower interspaces. The direction of the fibers of the external intercostal is the same as the direction of the external oblique muscles of the anterior abdominal wall, whereas the internal intercostal muscles are at roughly 90° to the direction of the external intercostals.

Figure 3.2   Anterior mediastinum. This anterior view of structures passing between the neck and the thorax within the circle of the first ribs displays clinically important anatomical structures. Most superficial is the thymus gland. Just beneath this are the major veins overlying major arterial trunks to the head, neck, and upper limb. Note the thyroid gland surrounding the trachea superiorly and the pericardium covering the heart and root of the great vessels inferiorly. Arteries and nerves to the upper limb pass behind the scalenus anterior muscle, as shown on the right. On the left, one sees the subclavian vein coursing over the first rib in front of the scalenus anterior muscle.

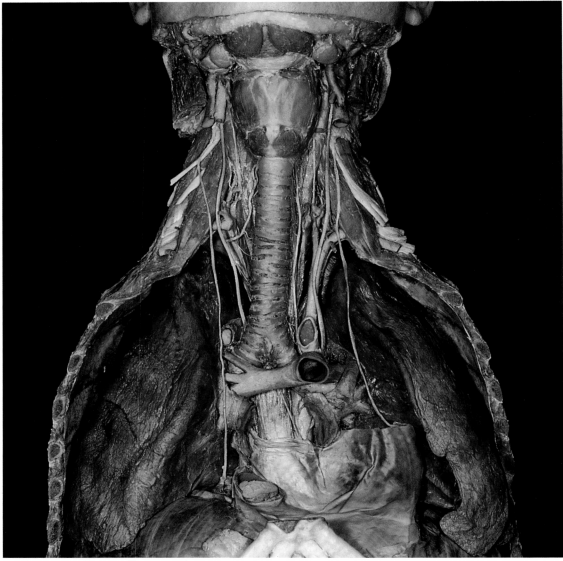

Figure 3.3A   Chest wall removed.

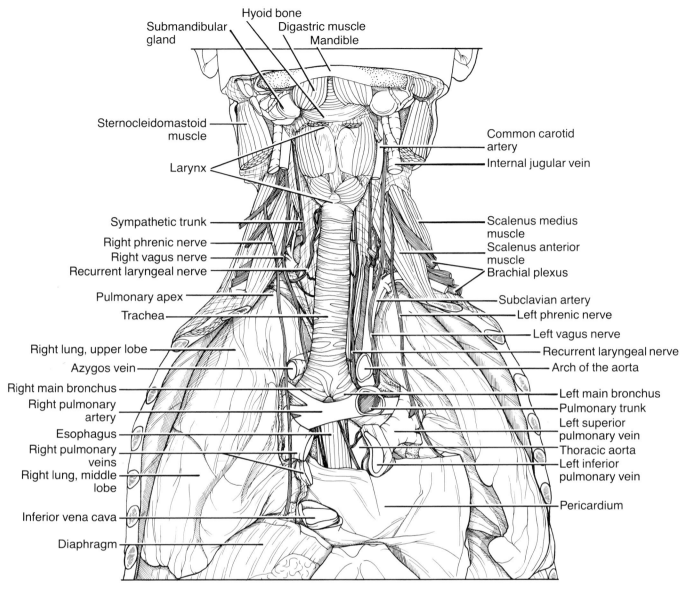

Hyoid bone
Submandibular gland
Digastric muscle
Mandible

Sternocleidomastoid muscle

Common carotid artery
Internal jugular vein

Larynx

Sympathetic trunk
Right phrenic nerve
Right vagus nerve
Recurrent laryngeal nerve

Scalenus medius muscle
Scalenus anterior muscle
Brachial plexus

Pulmonary apex
Trachea

Subclavian artery
Left phrenic nerve
Left vagus nerve

Right lung, upper lobe
Azygos vein
Right main bronchus
Right pulmonary artery
Esophagus
Right pulmonary veins
Right lung, middle lobe

Recurrent laryngeal nerve
Arch of the aorta
Left main bronchus
Pulmonary trunk
Left superior pulmonary vein
Thoracic aorta
Left inferior pulmonary vein

Inferior vena cava
Diaphragm

Pericardium

Figure 3.3B Chest wall removed.

Figure 3.3A, B Chest wall removed. The anterior part of the thoracic wall has been removed in this view of the larynx, trachea, main bronchi, and lungs. The relationship of the trachea and mainstem bronchi to the pulmonary arteries is clearly shown. The azygos vein on the right and the arch of the aorta on the left curve over the mainstem bronchi to gain access to the posterior mediastinal area. The lungs are in situ; on the right, the right upper and middle lobes may be seen. On the left, only the upper lobe is prominent. The phrenic nerves pass straight through the chest on the lateral walls of the mediastinum to innervate the right and left diaphragms.

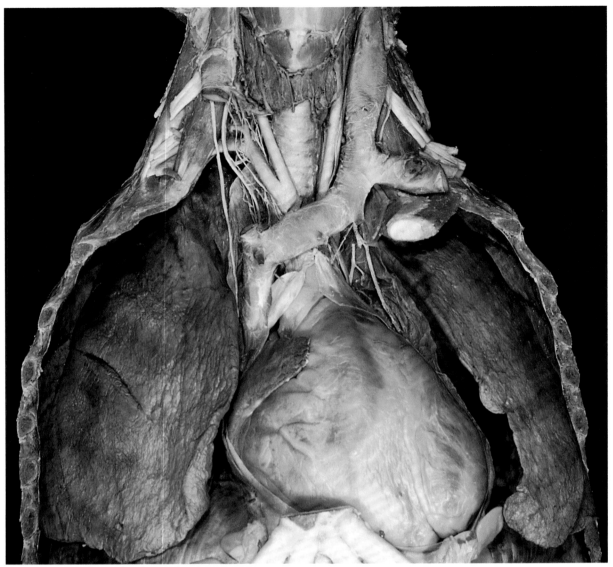

Figure 3.4A   Thoracic viscera in situ.

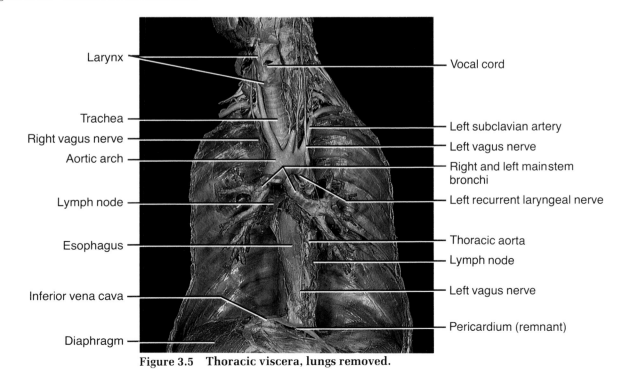

Larynx — Vocal cord

Trachea — Left subclavian artery

Right vagus nerve — Left vagus nerve

Aortic arch — Right and left mainstem bronchi

Lymph node — Left recurrent laryngeal nerve

Esophagus — Thoracic aorta

— Lymph node

Inferior vena cava — Left vagus nerve

— Pericardium (remnant)

Diaphragm —

Figure 3.5   Thoracic viscera, lungs removed.

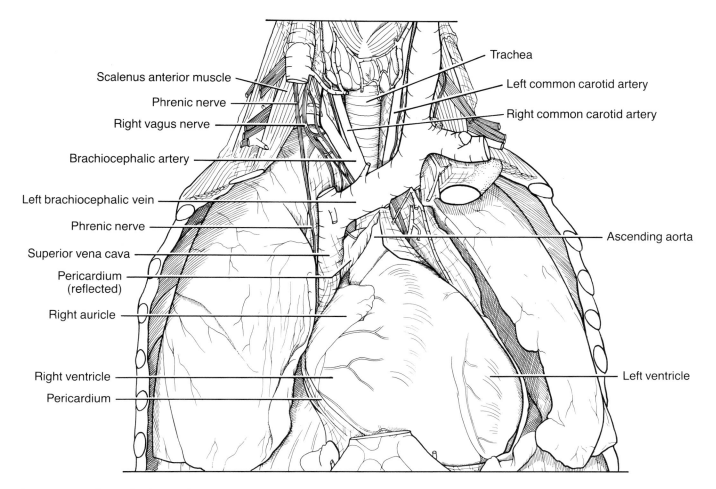

Scalenus anterior muscle

Phrenic nerve

Right vagus nerve

Brachiocephalic artery

Left brachiocephalic vein

Phrenic nerve

Superior vena cava

Pericardium (reflected)

Right auricle

Right ventricle

Pericardium

Trachea

Left common carotid artery

Right common carotid artery

Ascending aorta

Left ventricle

Figure 3.4B   Thoracic viscera in situ.

Figure 3.4A, B   Thoracic viscera in situ. This anterior view of the heart and great vessels in situ with the thymus gland removed shows the origin of the ascending aorta, still within the pericardial sac. The trachea and its relationships to the great vessels are evident in the superior midline. The important phrenic and vagus nerves pass through the inlet. The phrenic nerve is shown in relationship to the scalenus anterior and the vagus nerve in relationship to the carotid and major arterial trunks.

Figure 3.5   Thoracic viscera, lungs removed. This is an anterior view of the chest with the tracheal bronchial tree, aorta, and esophagus exposed. The arch of the aorta courses from anterior to posterior to the other structures as it arches from right to left to occupy its posterior position in the inferior chest. The right mainstem bronchus courses more in line with the trachea than does the longer left mainstem bronchus. The left recurrent laryngeal nerve courses around the arch of the aorta from anterior to posterior while the right arches behind the right subclavian artery.

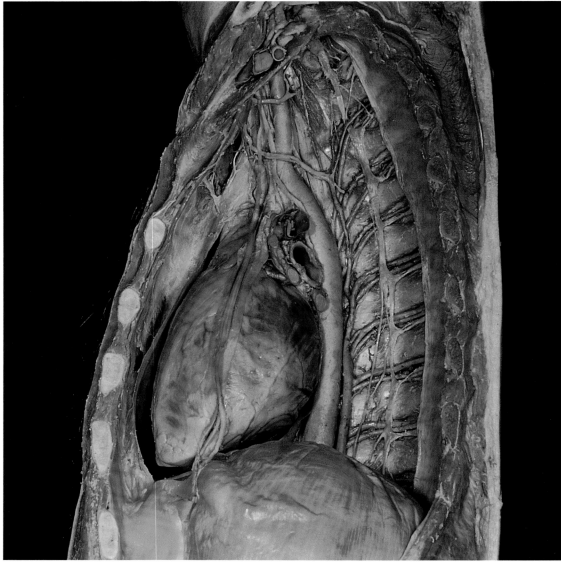

Figure 3.6A Lateral view of mediastinum.

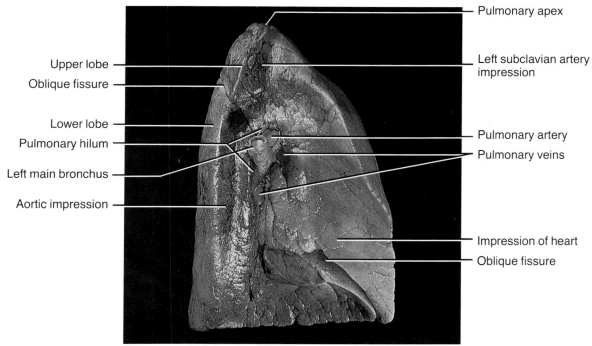

Figure 3.7 Left lung, mediastinal surface.

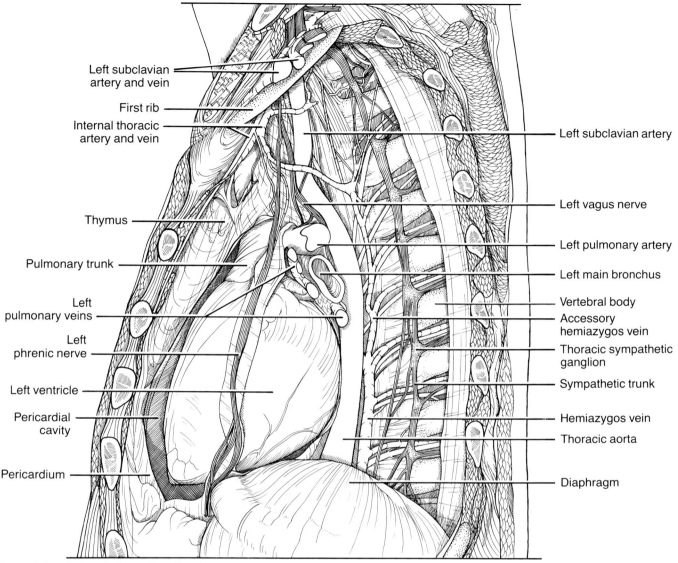

Left subclavian artery and vein

First rib

Internal thoracic artery and vein

Thymus

Pulmonary trunk

Left pulmonary veins

Left phrenic nerve

Left ventricle

Pericardial cavity

Pericardium

Left subclavian artery

Left vagus nerve

Left pulmonary artery

Left main bronchus

Vertebral body

Accessory hemiazygos vein

Thoracic sympathetic ganglion

Sympathetic trunk

Hemiazygos vein

Thoracic aorta

Diaphragm

Figure 3.6B   Lateral view of mediastinum.

Figure 3.6A, B   Lateral view of mediastinum. A view of the left side of the mediastinum with the pleura removed displays the relationships of the heart and great vessels to the vertebral column behind and the thymus and sternum in front. Structures to the upper limb are visible coursing over the curve of the first rib. The sympathetic chain and ganglia course along the anterolateral surface of the vertebral bodies. The pulmonary hilum, consisting of the left main bronchus, left pulmonary artery, and left pulmonary veins, is located in the mid-thorax, serving as the root of the removed left lung. The phrenic nerve courses through the chest on the pleural surface of the mediastinum from thoracic inlet to diaphragm. Just beneath the costal cartilages anteriorly, the internal thoracic artery and vein descend in close relationship to the thymus gland. The dome of the diaphragm appears at the level of the origin of the ninth rib.

Figure 3.7   Left lung, mediastinal surface. The medial surface of the left lung displays the lung hilum, consisting of the left main bronchus, the left pulmonary artery, and the pulmonary veins leaving the left lung. Note the division of the lung into an upper and lower lobe along the oblique fissure. The impression of the descending aorta and its arch and the subclavian artery are readily visible.

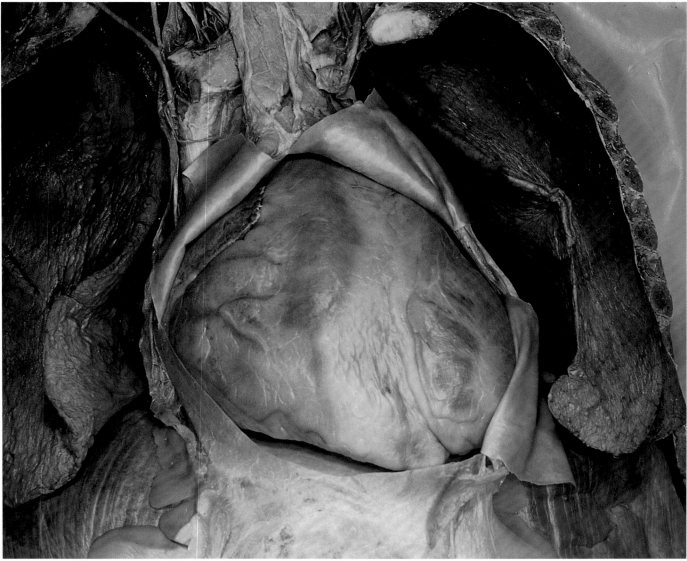

Figure 3.8A   Heart and pericardium.

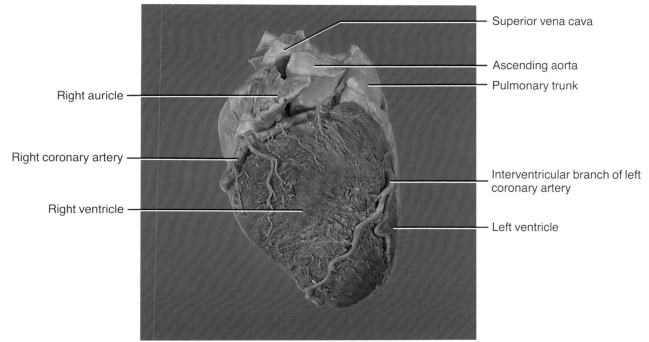

Figure 3.9   Heart and coronary arteries.

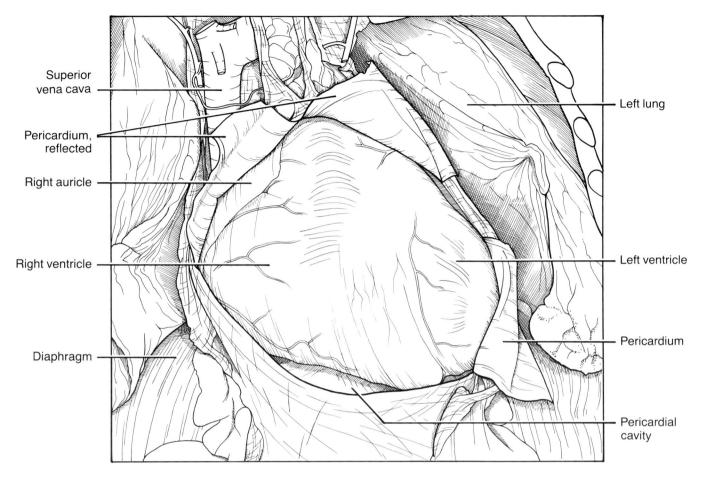

Superior vena cava

Pericardium, reflected

Right auricle

Right ventricle

Diaphragm

Left lung

Left ventricle

Pericardium

Pericardial cavity

Figure 3.8B   Heart and pericardium.

Figure 3.8A, B   Heart and pericardium. The heart is exposed within the pericardial cavity as the pericardial sac has been incised and its cut edges reflected. The visceral pericardium or epicardium over the surface of the heart remains intact.

Figure 3.9   Heart and coronary arteries. The anterior view of the heart shows the relationships of the great vessels at the root of the heart, its major chambers, and the coronary arteries on the anterior surface. The right auricle and ventricle make up the major anterior surface of the heart; the left ventricle constitutes its right border and apex. The right coronary artery and the interventricular branch of the left coronary artery are visible on the anterior surface of the heart.

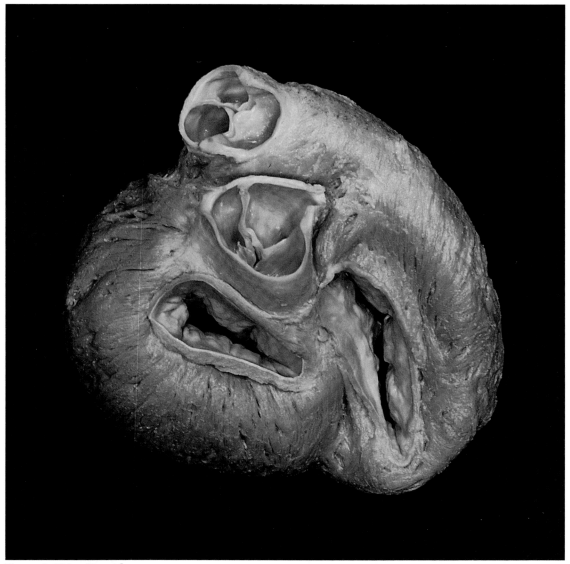

Figure 3.10A   Heart base.

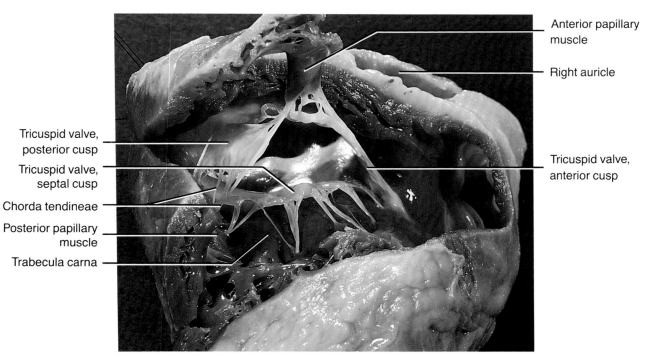

Anterior papillary muscle

Right auricle

Tricuspid valve, posterior cusp

Tricuspid valve, septal cusp

Chorda tendineae

Posterior papillary muscle

Trabecula carna

Tricuspid valve, anterior cusp

Figure 3.11   Heart valve.

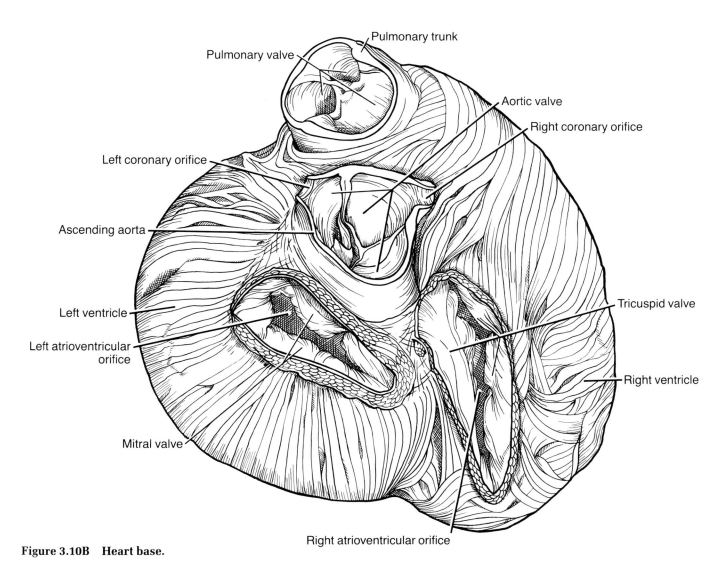

Figure 3.10B   Heart base.

Figure 3.10A, B   Heart base. A view down on the superior surface of the dissected heart shows the relationship of the great vessels at the root of the heart. With the right and left atrium open, the tricuspid and mitral valves are visible. The cusps of the pulmonary and aortic valves are clear. Orifices to the left and right coronary arteries from right and left aortic sinuses have been cut off immediately after their exit from the aorta.

Figure 3.11   Heart valve. A dissection of the tricuspid heart valve viewed from below displays the chorda tendinae to the edges of the valve leaflets from the papillary muscles. The interior of the right atrium is visible through the ostium of the valve.

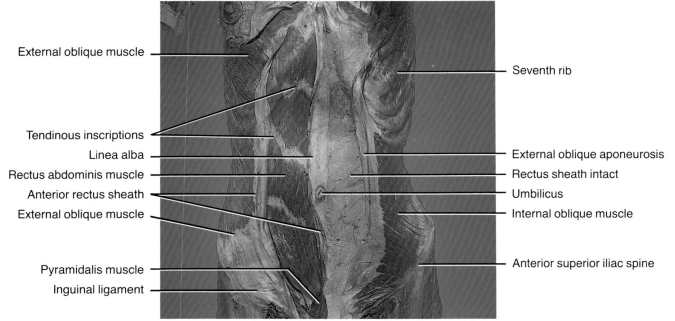

External oblique muscle

Tendinous inscriptions

Linea alba

Rectus abdominis muscle

Anterior rectus sheath

External oblique muscle

Pyramidalis muscle

Inguinal ligament

Seventh rib

External oblique aponeurosis

Rectus sheath intact

Umbilicus

Internal oblique muscle

Anterior superior iliac spine

Figure 4.1   Abdominal wall, rectus uncovered.

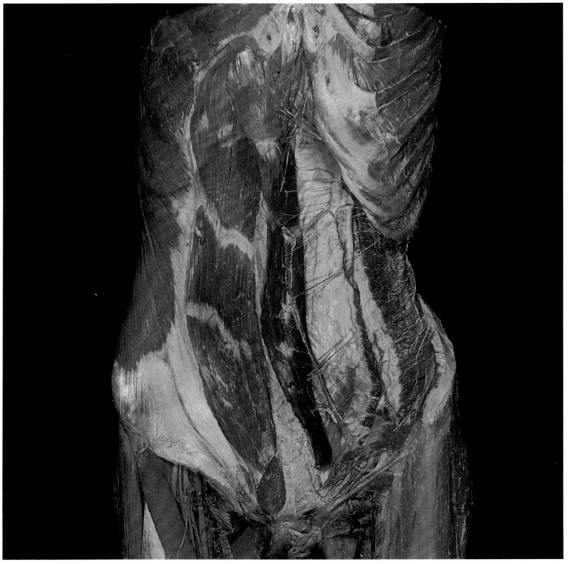

Figure 4.2A   Abdominal wall, rectus reflected.

Figure 4.1 Abdominal wall, rectus uncovered. This view of the anterior abdominal wall with the right rectus sheath open exposes the rectus muscle, displaying its tendinous inscriptions. On the right, the external oblique muscle is left in situ. On the left, it is removed to expose the internal oblique muscle. The cut edge of the external oblique aponeurosis participates in the formation of the intact anterior rectus sheath on the left.

Figure 4.2A, B Abdominal wall, rectus reflected. With the anterior rectus sheath removed on the right and the rectus muscle cut proximally and reflected to the right, this general view of the anterior abdominal wall shows the course of nerves and blood vessels to this important muscle. The superior and inferior epigastric vessels join on the deep surface of the rectus muscle. The intercostal nerves that innervate the rectus muscle course to it on the surface of the transversus abdominis muscle deep to both the external and internal oblique muscles. They penetrate the sheath of the rectus abdominis muscle to give it motor supply.

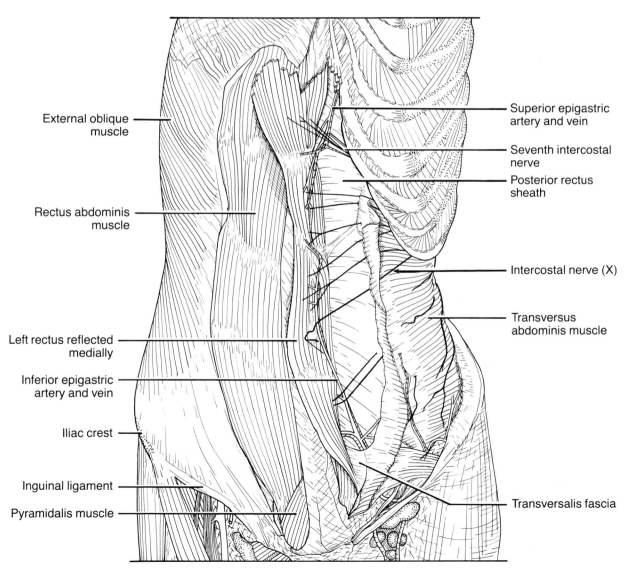

Figure 4.2B Abdominal wall, rectus reflected.

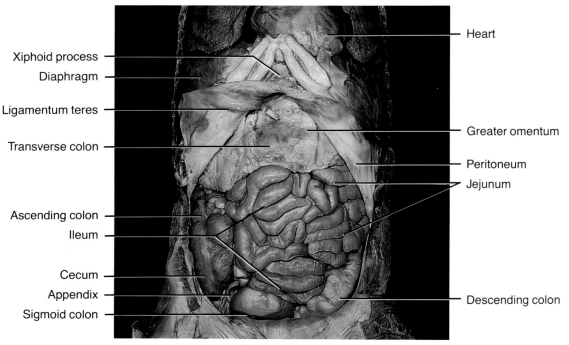

Xiphoid process

Diaphragm

Ligamentum teres

Transverse colon

Ascending colon

Ileum

Cecum

Appendix

Sigmoid colon

Heart

Greater omentum

Peritoneum

Jejunum

Descending colon

Figure 4.3   Abdominal organs in situ, omentum up.

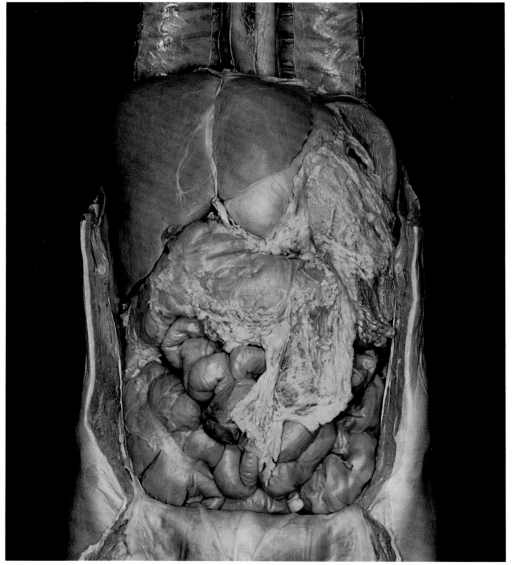

Figure 4.4A ˜ Abdominal organs in situ.

Figure 4.3   Abdominal organs in situ, omentum up. The peritoneal cavity is open by reflecting the abdominal wall and its peritoneum. The greater omentum has been reflected upward to expose the loops of the jejunum and ileum. The ascending colon, cecum, and appendix lie in the right extreme portion of the peritoneal cavity. The descending colon is visible on the left. The transverse colon, seen through the omentum, frames the superior part of the small bowel compartment. The sigmoid colon can be seen inferiorly. The upper abdominal contents, including the liver, stomach, spleen, and pancreas, are hidden by the reflected greater omentum.

Figure 4.4A, B   Abdominal organs in situ. The diaphragm and rib cage are removed to expose the upper abdominal contents. Note the liver, with its falciform ligament between its right and left lobes. The falciform ligament terminates as the ligamentum teres. The relationship of liver and spleen to the cut edge of the diaphragm is evident. The greater omentum is draped downward in normal position to show the transverse colon. The relationship of the gall bladder on the inferior surface of the liver to the colon is also evident. The stomach lies obliquely along the inferior surface of the left lobe of the liver, extending upward to the level of the diaphragm and downward to a variable distance according to the quantity of its contents.

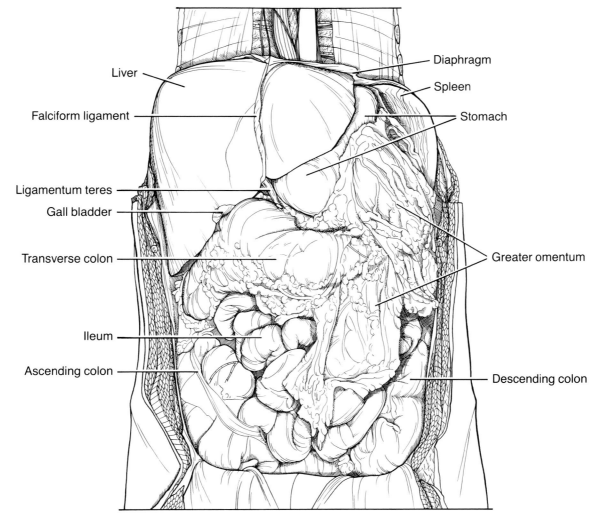

Figure 4.4B   Abdominal organs in situ.

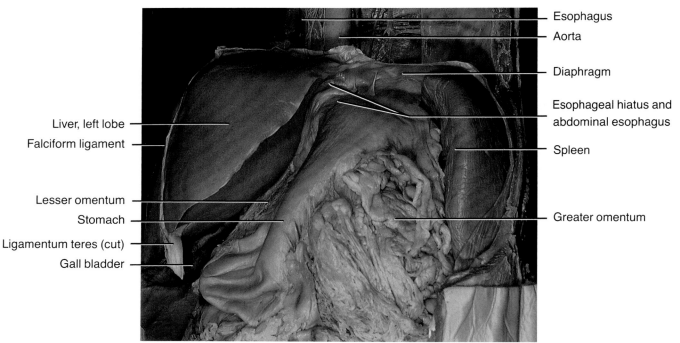

Esophagus

Aorta

Diaphragm

Esophageal hiatus and abdominal esophagus

Spleen

Liver, left lobe

Falciform ligament

Lesser omentum

Stomach

Ligamentum teres (cut)

Gall bladder

Greater omentum

Figure 4.5   Stomach, liver, and spleen in situ.

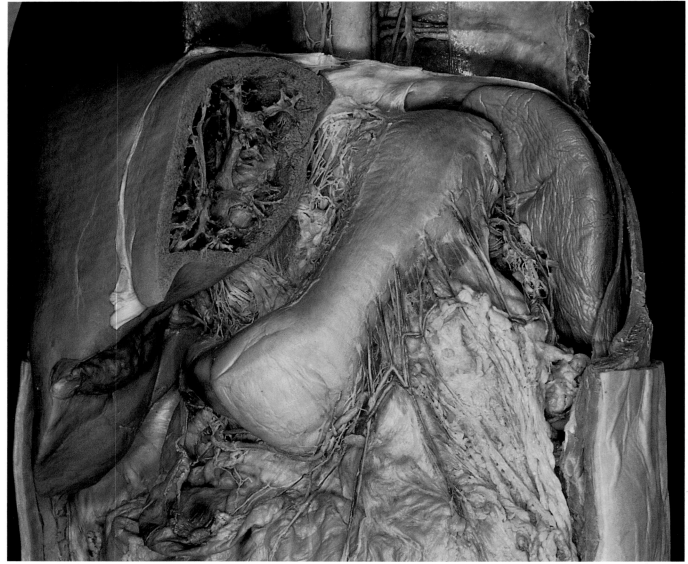

Figure 4.6A   Stomach, liver, and spleen, partial dissection.

Figure 4.5 Stomach, liver, and spleen in situ. The relationships of the liver, stomach, diaphragm, and spleen are shown here with the rib cage removed. The esophagus penetrates the diaphragm to the right and anterior to the aorta. The lesser omentum extends from the lesser curvature of the stomach to the liver. The origin of the greater omentum is seen from the greater curvature of the stomach. It drapes over the transverse colon and forms an apron down the front of the peritoneal cavity. The spleen is subdiaphragmatic, lying protected by the inferior left rib cage.

Figure 4.6A, B Stomach, liver, and spleen, partial dissection. The left lobe of the liver has been sectioned to show some of its internal structure and the relationship of the liver to the stomach. The liver and lesser curvature of the stomach are connected by the lesser omentum. The fundus, body, pyloric antrum, and pylorus of the stomach are identified. The fundus of the stomach is cradled by the medial surface of the spleen. The greater curvature gives rise to the greater omentum, in which one may see the gastroepiploic blood vessels supplying the greater curvature and omentum itself. The common bile duct is dissected from the peritoneum of the lesser omentum to show its relationship to the liver and duodenum beyond the pylorus.

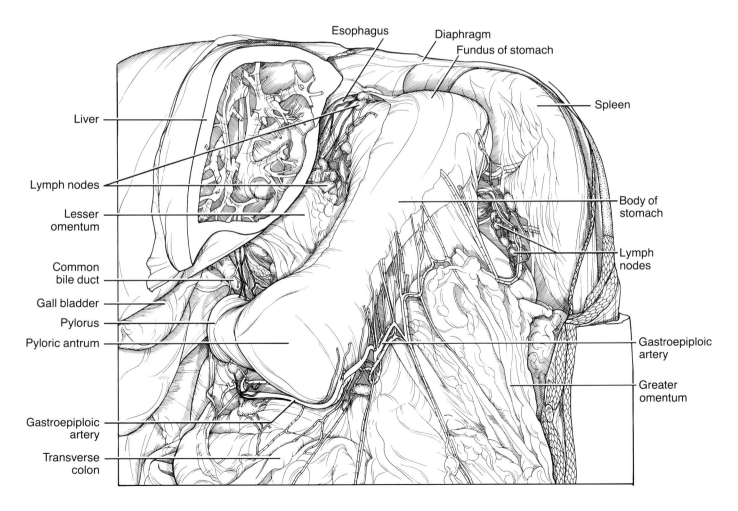

Figure 4.6B Stomach, liver, and spleen, partial dissection.

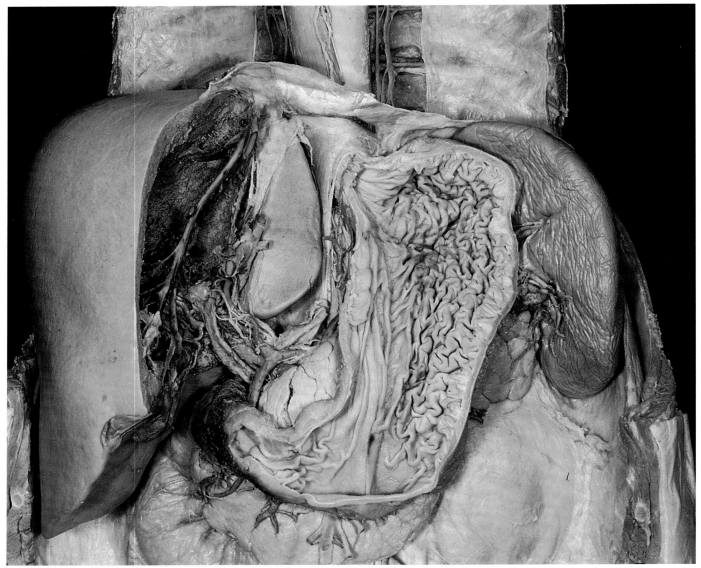

Figure 4.7A   Stomach, open.

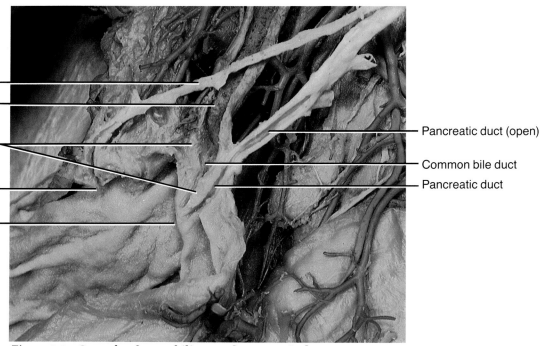

Accessory pancreatic duct

Common bile duct (open)

Hepatopancreatic sphincter (Oddi)

Accessory pancreatic duct orifice

Major duodenal papilla (at Ampulla of Vater)

Pancreatic duct (open)

Common bile duct

Pancreatic duct

Figure 4.8   Open duodenum biliary and pancreatic ducts.

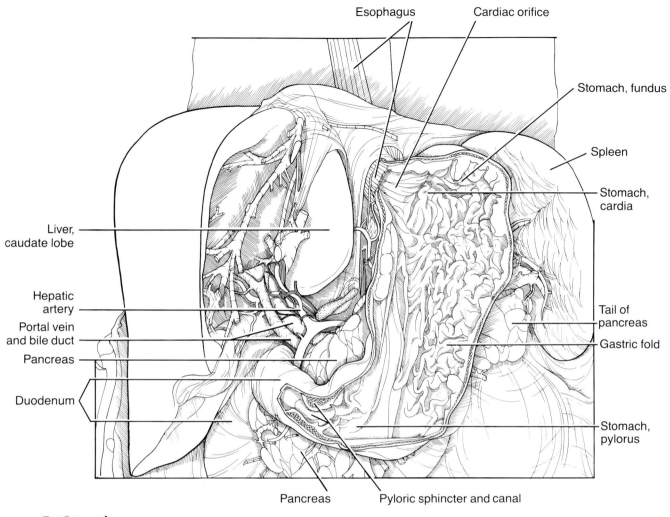

Figure 4.7B   Stomach, open.

Figure 4.7A, B   Stomach, open. The stomach has been opened on its anterior surface to expose the mucosal lining. The characteristics of the cardia and cardiac orifice, the fundus, and the pylorus are displayed. The cut edge of the pyloric sphincter surrounding the pyloric canal appears at the entry into the first portion of the duodenum. Relationships of the hepatic artery, portal vein, and common bile duct to one another and to the pancreas and duodenum are evident in this dissection. Peeking from beneath the greater curvature of the stomach is the tail of the pancreas as it extends toward the hilum of the spleen.

Figure 4.8   Open duodenum biliary and pancreatic ducts. The anterior surface of the duodenum has been cut away to show the entry of the pancreatico-biliary ducts through the duodenal wall. The common bile duct and pancreatic duct open separately at the duodenal papilla in this specimen, although frequently they join before traversing the duodenal wall. The accessory pancreatic duct has its own orifice into the duodenum proximal to the major duodenal papilla at the ampulla for the common bile duct and pancreatic duct (Ampulla of Vater).

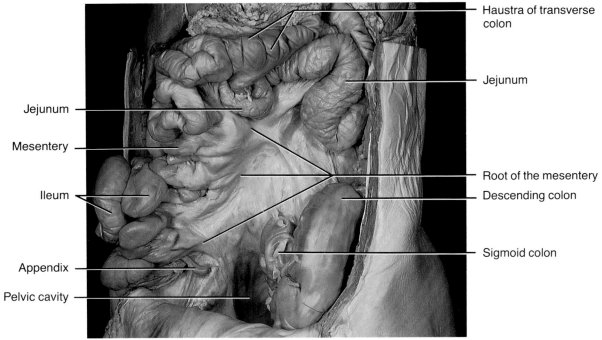

Haustra of transverse colon

Jejunum

Jejunum

Mesentery

Root of the mesentery

Descending colon

Ileum

Appendix

Sigmoid colon

Pelvic cavity

Figure 4.9   Small bowel mesentery.

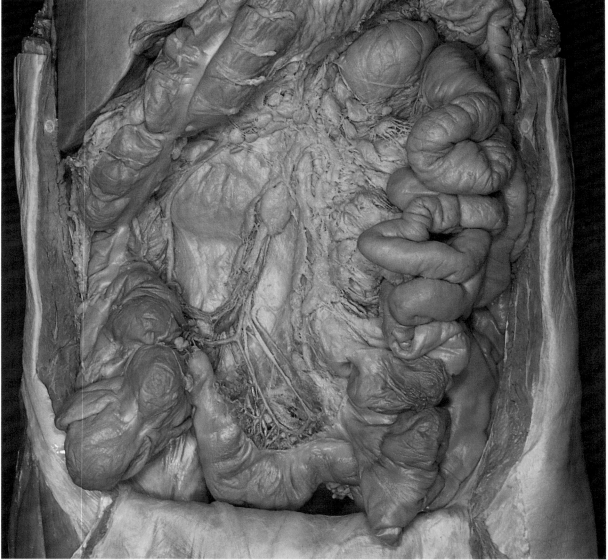

Figure 4.10A   Small bowel mesentery, dissected.

Figure 4.9 Small bowel mesentery. The small bowel, both jejunum and ileum, has been reflected upward to show the obliquely placed root of the mesentery, which carries its blood, lymph, and nerve supply. Retraction of the small bowel exposes the descending and sigmoid colon. The transverse colon is identifiable by its haustra and barely visible taenia on its inferior surface.

Figure 4.10A, B Small bowel mesentery, dissected. With the jejunum and ileum retracted to the left and the colon reflected upward, structures within the small bowel mesentery and mesocolon are visible. The peritoneum, which covered the surface of the mesentery, has been removed to show lymphatic structures and blood vessels.

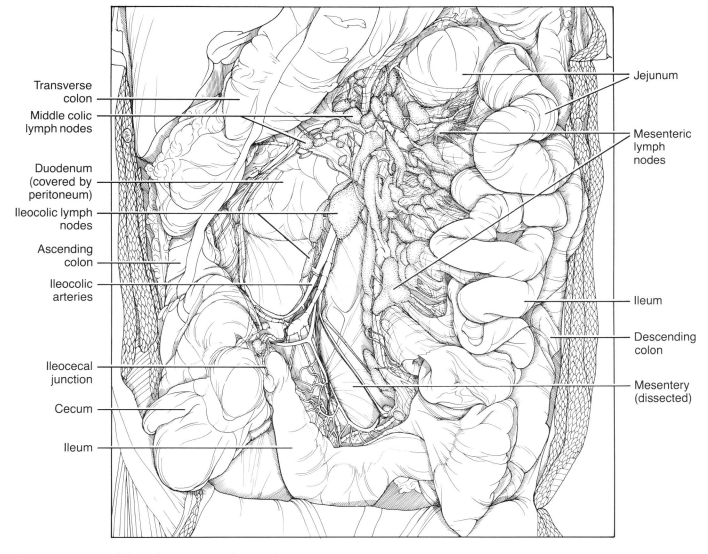

Figure 4.10B Small bowel mesentery, dissected.

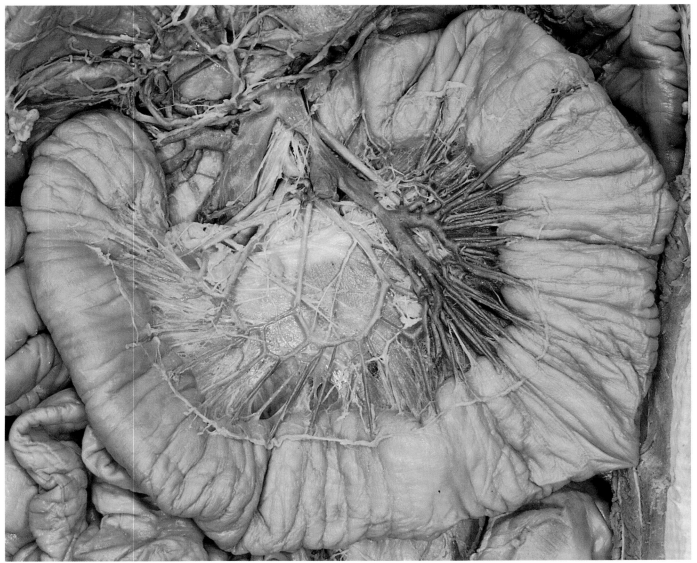

Figure 4.11A   Small bowel mesentery, dissected, detail.

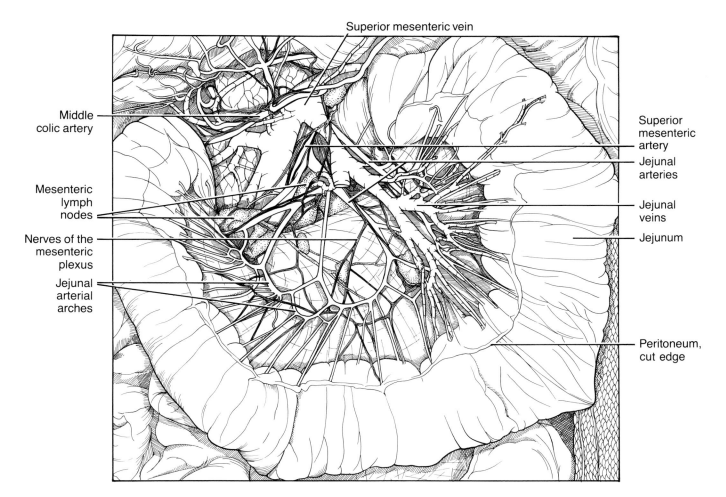

Superior mesenteric vein

Middle colic artery

Mesenteric lymph nodes

Nerves of the mesenteric plexus

Jejunal arterial arches

Superior mesenteric artery

Jejunal arteries

Jejunal veins

Jejunum

Peritoneum, cut edge

Figure 4.11B   Small bowel mesentery, dissected, detail.

Figure 4.11A, B   Small bowel mesentery, dissected, detail. A loop of jejunum has been isolated, and the peritoneum has been removed from one surface of its mesentery. The vessels and nerves approaching the intestinal wall are readily visible. Note the jejunal arterial arches and the straight arterial branches to the jejunal wall from these arches. Lymph nodes are prominent in the jejunal mesentery.

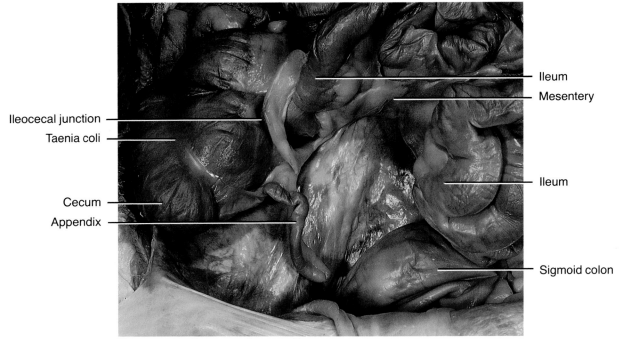

Ileum

Mesentery

Ileocecal junction

Taenia coli

Ileum

Cecum

Appendix

Sigmoid colon

Figure 4.12   Cecum and appendix.

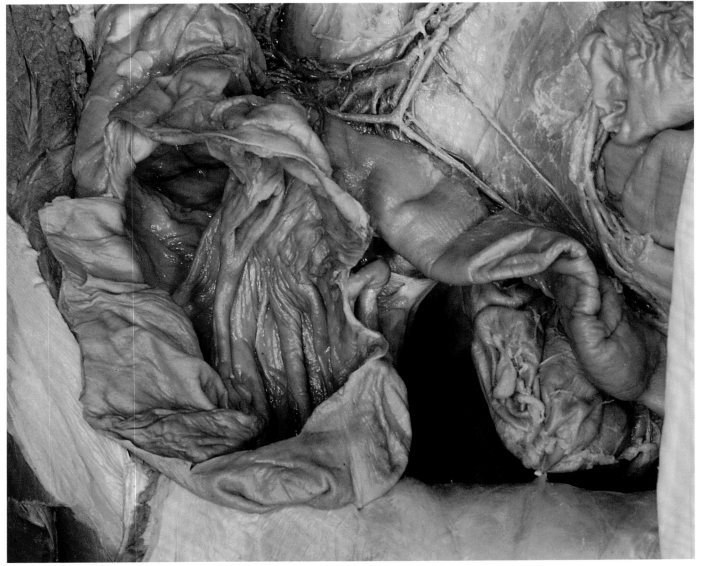

Figure 4.13A   Cecum, open.

Figure 4.12 **Cecum and appendix.** A view of the right lower quadrant of the abdominal cavity shows the junction of ileum and cecum. The appendix lies at the dependent end of the cecum at the confluence of the taeniae coli.

Figure 4.13A, B **Cecum, open.** Viewing the interior of the cecum, one may see the ileocecal valve and orifice of entry from ileum to cecum. The orifice of the vermiform appendix is visible, and the internal aspect of the haustral folds can be seen in the ascending colon.

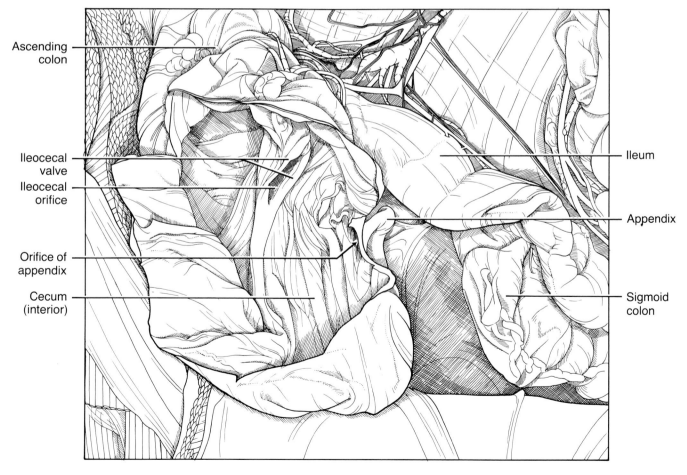

Ascending colon

Ileocecal valve

Ileocecal orifice

Orifice of appendix

Cecum (interior)

Ileum

Appendix

Sigmoid colon

Figure 4.13B **Cecum, open.**

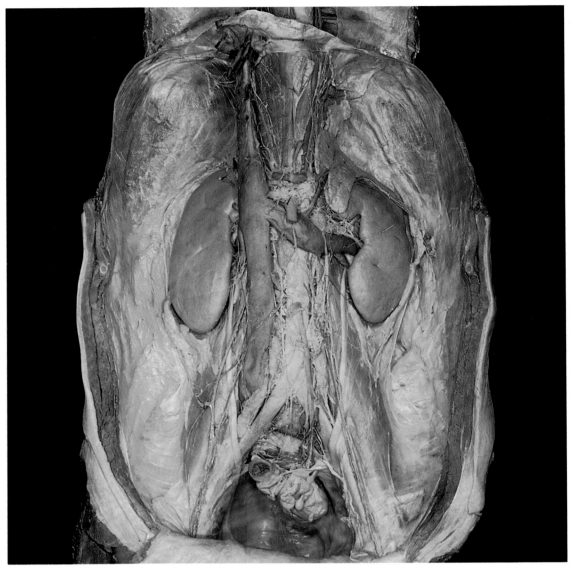

Figure 4.14A   Retroperitoneum.

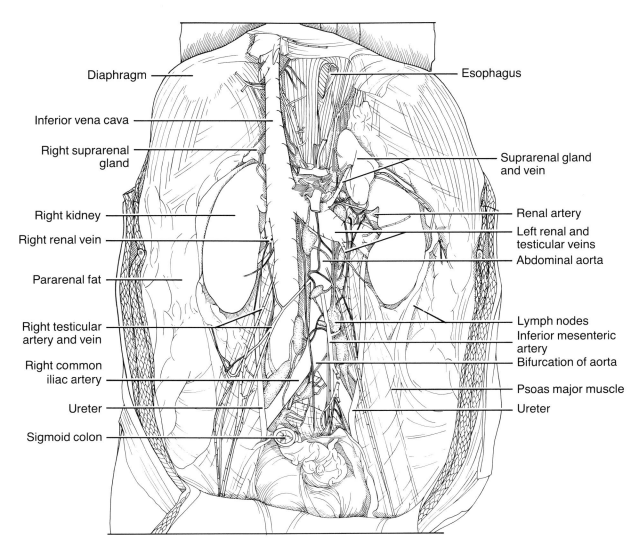

Diaphragm

Inferior vena cava

Right suprarenal gland

Right kidney

Right renal vein

Pararenal fat

Right testicular artery and vein

Right common iliac artery

Ureter

Sigmoid colon

Esophagus

Suprarenal gland and vein

Renal artery

Left renal and testicular veins

Abdominal aorta

Lymph nodes

Inferior mesenteric artery

Bifurcation of aorta

Psoas major muscle

Ureter

Figure 4.14B   Retroperitoneum.

Figure 4.14A, B   Retroperitoneum. Retroperitoneal structures are viewed by removal of the anterior layer of the renal fascia and posterior parietal peritoneum. This exposes the ureters, testicular vessels, kidneys, suprarenal glands, aorta, and vena cava with their associated branches and lymphatic structures and nerve plexuses. The vena cava is to the right of the aorta, and the left renal vein crosses superficial to it to reach the left kidney. Note the left suprarenal gland with its vein emptying into the left renal vein. The ureters descend on the psoas muscles to cross superficial to the great vessels as they enter the true pelvis to empty into the posterior wall of the bladder. The right and left testicular arteries arise from the aorta. The right testicular vein empties into the vena cava, but the left testicular vein empties into the left renal vein.

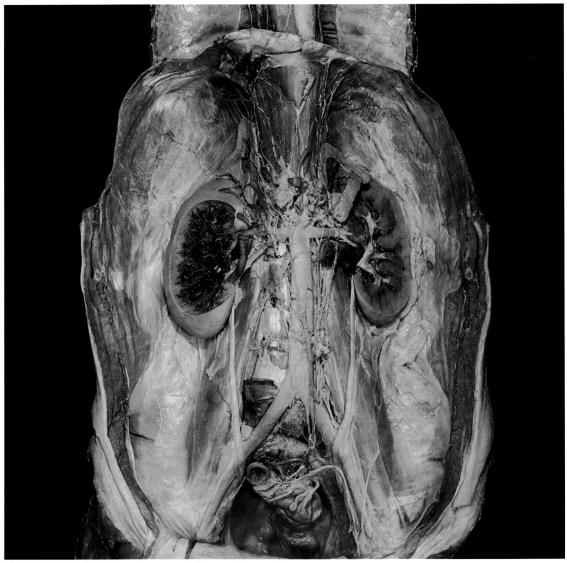

Figure 4.15A   Retroperitoneum, kidneys dissected.

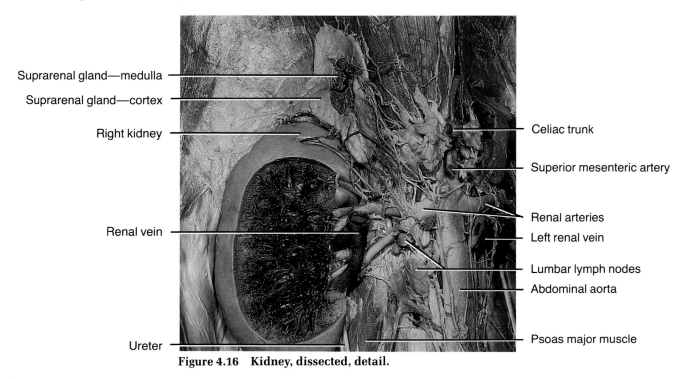

Suprarenal gland—medulla

Suprarenal gland—cortex

Right kidney

Renal vein

Ureter

Celiac trunk

Superior mesenteric artery

Renal arteries

Left renal vein

Lumbar lymph nodes

Abdominal aorta

Psoas major muscle

Figure 4.16   Kidney, dissected, detail.

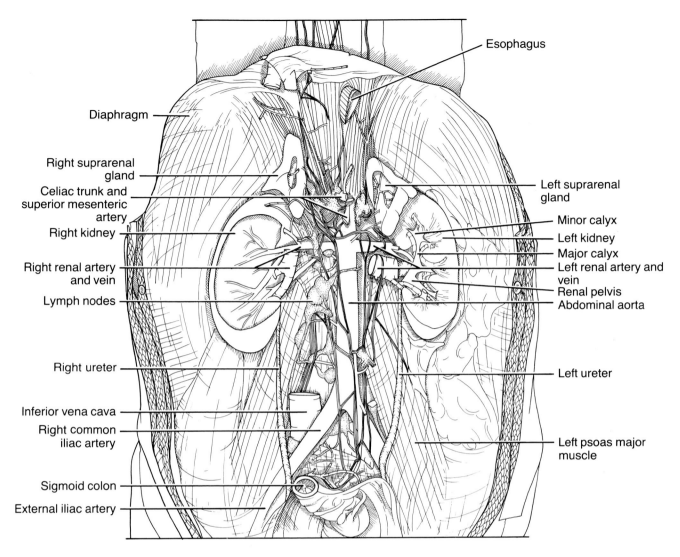

Esophagus

Diaphragm

Right suprarenal gland

Celiac trunk and superior mesenteric artery

Right kidney

Right renal artery and vein

Lymph nodes

Right ureter

Inferior vena cava

Right common iliac artery

Sigmoid colon

External iliac artery

Left suprarenal gland

Minor calyx

Left kidney

Major calyx

Left renal artery and vein

Renal pelvis

Abdominal aorta

Left ureter

Left psoas major muscle

Figure 4.15B   Retroperitoneum, kidneys dissected.

Figure 4.15A, B   Retroperitoneum, kidneys dissected. The right kidney has been dissected with removal of its anterior surface to demonstrate the distribution of renal blood vessels. The left kidney has been sectioned longitudinally to display the vessels and the renal pelvis with its major and minor calyces. The central area of each suprarenal gland has been dissected and removed to expose the tributaries of the suprarenal vein. Note the celiac trunk and superior mesenteric artery arising above the renal vein and renal arteries to supply blood to the foregut.

Figure 4.16   Kidney, dissected, detail. Dissection of the renal parenchyma has exposed the (latex-filled) arteries and veins within the kidney. Smaller vessels have been trimmed away. The cortex of the suprarenal gland has been removed near its center to expose the brownish medullary tissue and to demonstrate the tributaries to the suprarenal veins from the medulla.

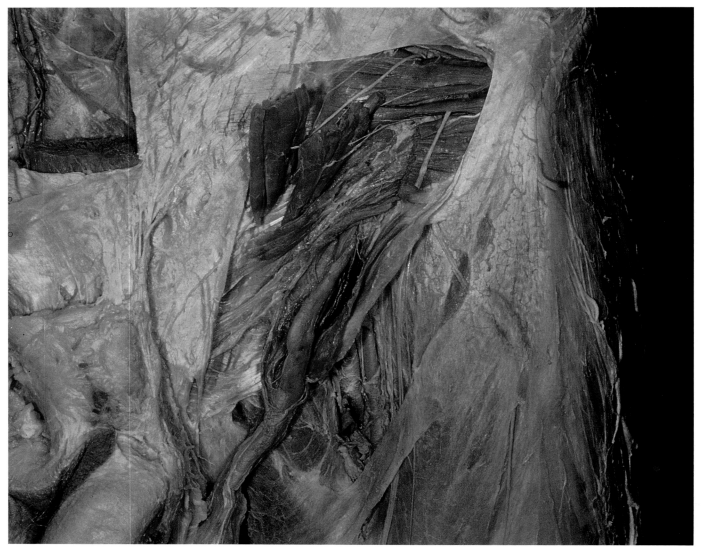

Figure 5.1A   Left inguinal area.

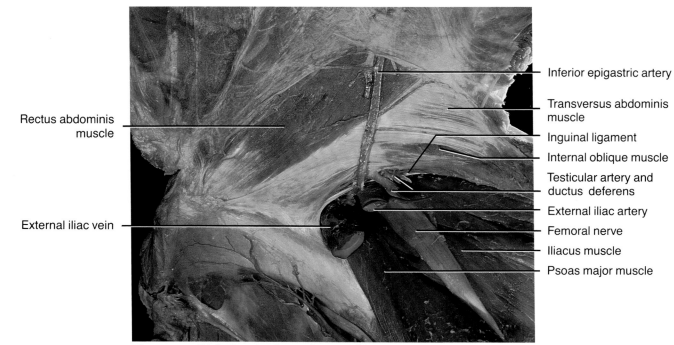

Rectus abdominis muscle

External iliac vein

Inferior epigastric artery

Transversus abdominis muscle

Inguinal ligament

Internal oblique muscle

Testicular artery and ductus deferens

External iliac artery

Femoral nerve

Iliacus muscle

Psoas major muscle

Figure 5.2   Right inguinal area from within.

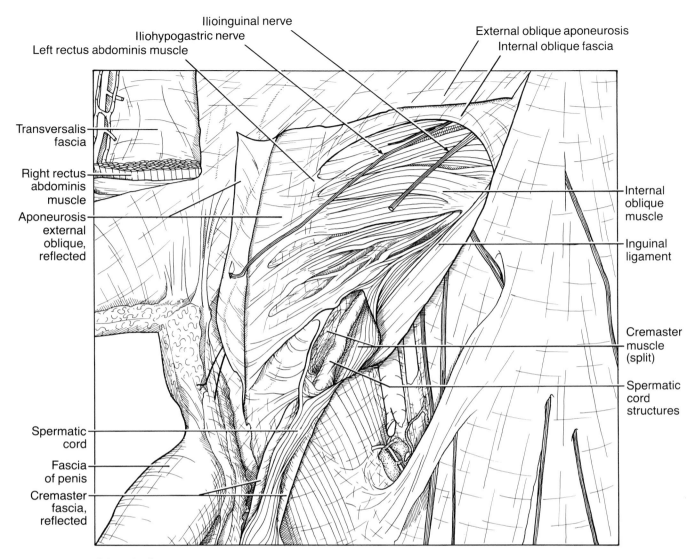

Ilioinguinal nerve

Iliohypogastric nerve

Left rectus abdominis muscle

External oblique aponeurosis

Internal oblique fascia

Transversalis fascia

Right rectus abdominis muscle

Aponeurosis external oblique, reflected

Internal oblique muscle

Inguinal ligament

Cremaster muscle (split)

Spermatic cord structures

Spermatic cord

Fascia of penis

Cremaster fascia, reflected

Figure 5.1B   Left inguinal area.

Figure 5.1A, B   Left inguinal area. Dissection of the male inguinal region's spermatic cord reveals the spermatic cord structures surrounded by the cremaster muscle passing along the superior surface of the inguinal ligament over its medial one-third. The external oblique has been reflected and the cremaster muscle split to show the spermatic cord structures within as they pass beneath the external oblique fascia to emerge through the external ring. The external ring has been sectioned in this dissection to expose structures descending into the scrotum.

Figure 5.2   Right inguinal area from within. A view from the inside shows structures passing both above and beneath the inguinal ligament to the scrotum and lower limb, respectively. The testicular vessels and ductus deferens pass through the internal inguinal ring superior to the inguinal ligament. Beneath the inguinal ligament, the external iliac artery and vein pass on the surface of the iliopsoas muscles. The inferior epigastric artery branches from the external iliac on the deep side of the abdominal wall just medial to the internal inguinal ring.

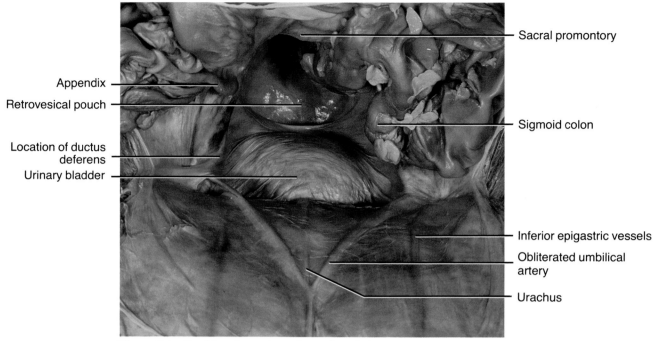

Appendix

Retrovesical pouch

Location of ductus
deferens

Urinary bladder

Sacral promontory

Sigmoid colon

Inferior epigastric vessels

Obliterated umbilical
artery

Urachus

Figure 5.3   Looking down into the male pelvis.

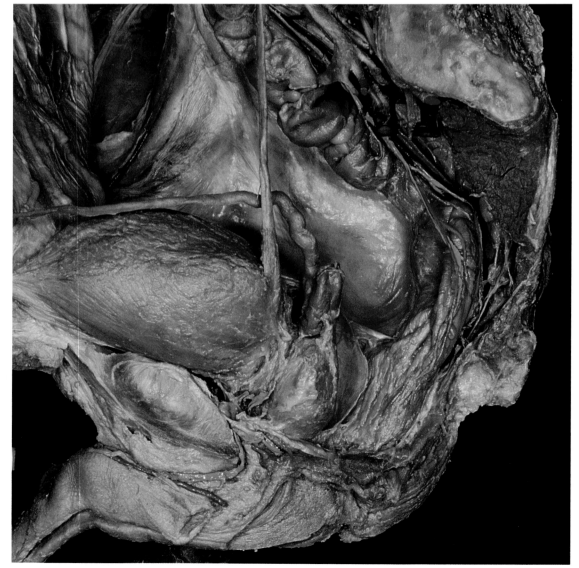

Figure 5.4A   Sagittal section, male pelvis.

Figure 5.3 Looking down into the male pelvis. A view downward into the male pelvis with the lower abdominal wall and its peritoneum left intact. The midline obliterated urachus and the folds for the two obliterated umbilical arteries meeting at the umbilicus lie just beneath the abdominal wall peritoneum, as do the inferior epigastric artery and vein. The appendix is visible in the false pelvis just above the commodious retrovesical pouch of the true pelvis.

Figure 5.4A, B Sagittal section, male pelvis. The male pelvis, dissected from a left lateral approach, shows relations of the bladder, the prostate, and left seminal vesicle to the rectum and the rectovesical pouch. The low posterior entry of the ureter into the bladder and the relationship of the left ureter to the left ductus deferens are clearly shown.

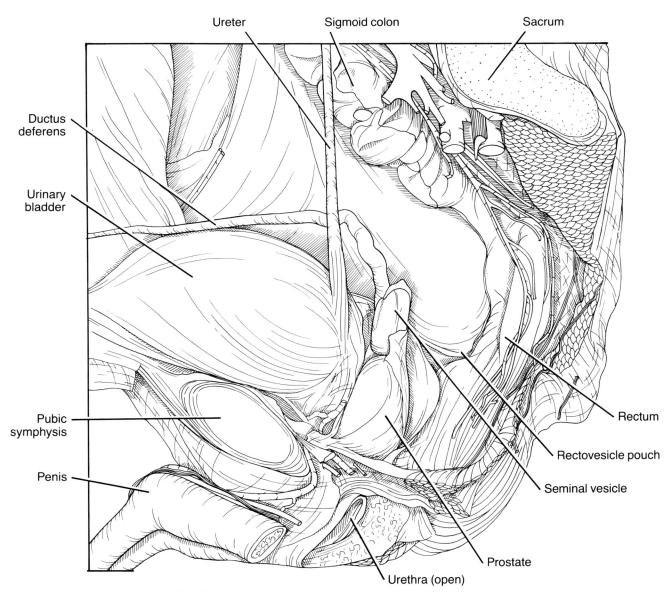

Figure 5.4B Sagittal section, male pelvis.

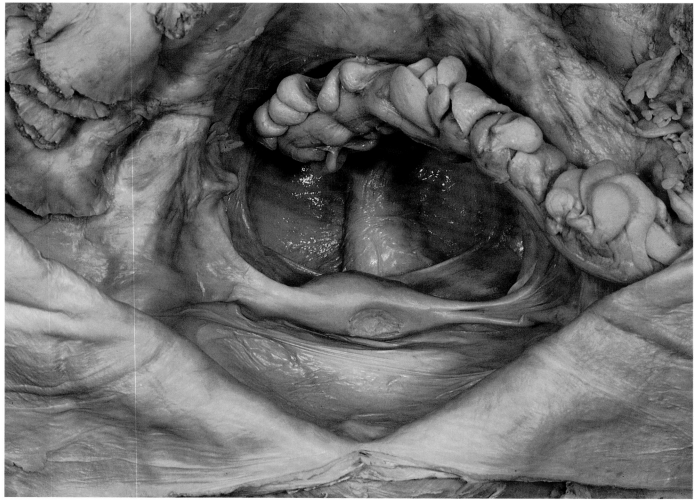

Figure 5.5A Looking down into the female pelvis.

Uterine tube (fimbriated end)   Sigmoid colon   Rectum

Uterine tube and ovary

Uterine tube

Round ligament of uterus

Broad ligament of uterus

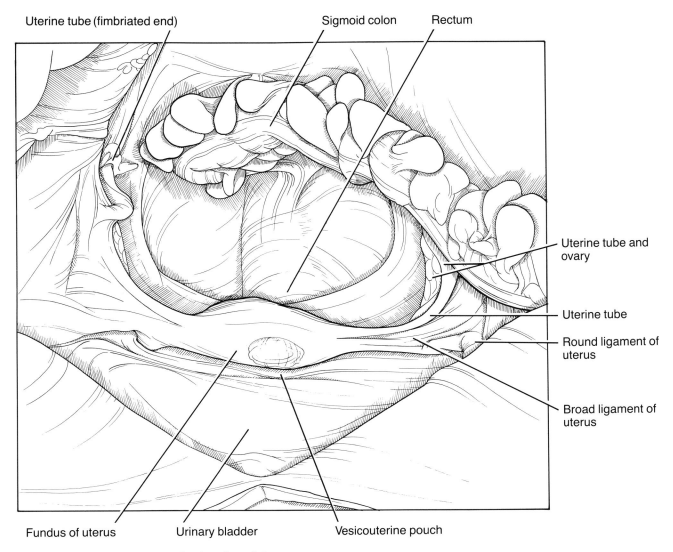

Fundus of uterus   Urinary bladder   Vesicouterine pouch

Figure 5.5B   Looking down into the female pelvis.

Figure 5.5A, B   Looking down in female pelvis. A view downward into the female pelvis displays the uterus anteriorly separated from the bladder by the peritoneally lined vesicouterine pouch. The deeper uterine pouch lies between the uterus and rectum to the floor of the pelvis. The oval depression in the fundus of the uterus is a site where a small fibroma was removed. The location of the uterine tube at the posterior side of the broad ligament and its relationship to the ovary within the true pelvis are visible.

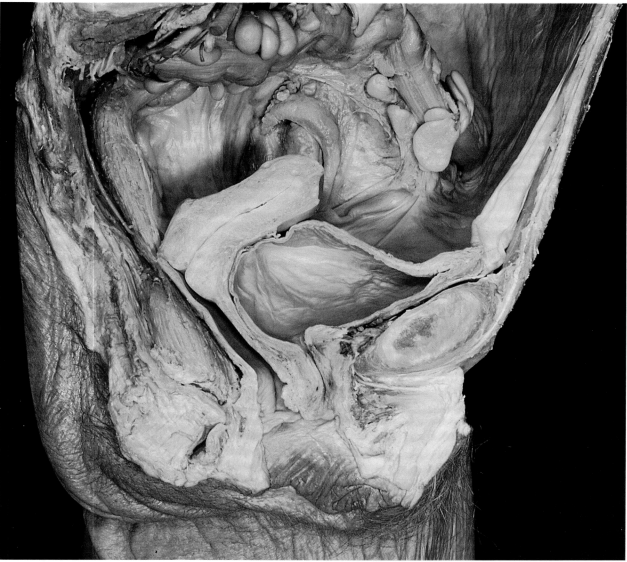

Figure 5.6A   Sagittal section, female pelvis.

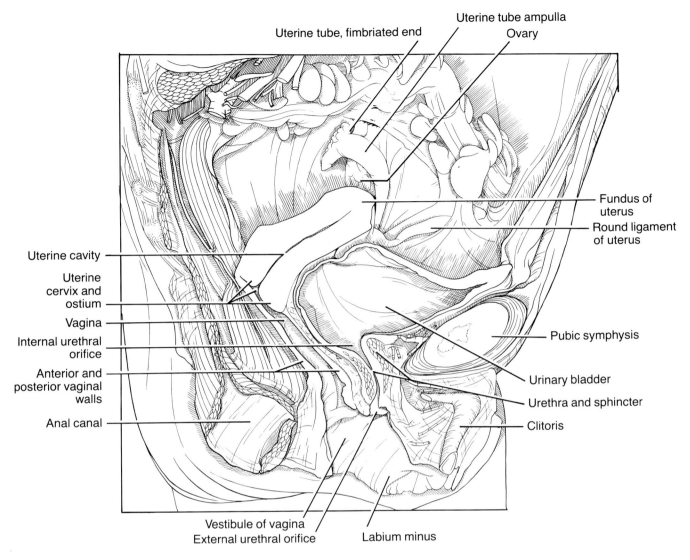

Uterine tube, fimbriated end
Uterine tube ampulla
Ovary
Fundus of uterus
Round ligament of uterus
Uterine cavity
Uterine cervix and ostium
Vagina
Internal urethral orifice
Anterior and posterior vaginal walls
Anal canal
Pubic symphysis
Urinary bladder
Urethra and sphincter
Clitoris
Vestibule of vagina
External urethral orifice
Labium minus

Figure 5.6B   Sagittal section, female pelvis.

Figure 5.6A, B   Sagittal section, female pelvis. A median section of the female pelvis displays the relationships of the bladder, urethra, uterus, vagina, and rectum. The urinary bladder behind the symphysis pubis is open to show its mucosal surface. The urethra and its sphincter are sectioned to show its relationship to the neck of the bladder and the external urethral orifice. The body of the uterus is tilted forward, placing the fundus along the posterior and superior surface of the bladder. The relationship of the anus and rectum to the pelvic structures is important in understanding what structures may be palpated in rectal examination. The relationships of clitoris, external urethral orifice, and vestibule of the vagina are clearly visible.

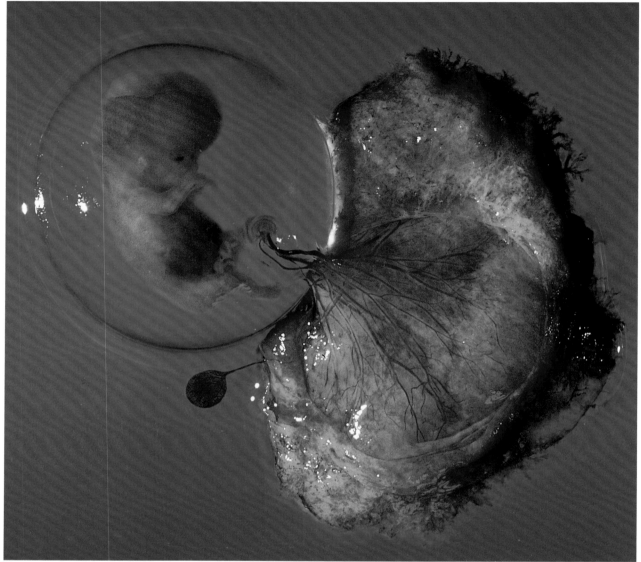

Figure 5.7A   Fetus and placenta.

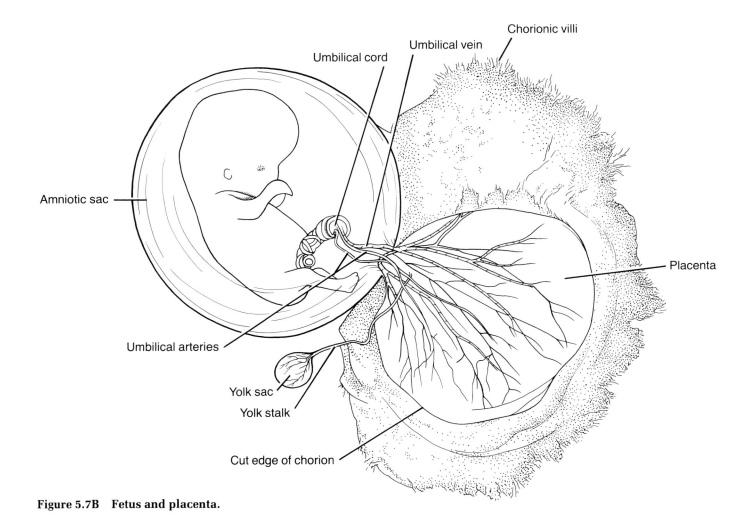

Figure 5.7B   Fetus and placenta.

Figure 5.7A, B   Fetus and placenta. A two-month-old embryo within its amniotic sac displays the placental and fetal elements. The yolk sac remains attached to the umbilical cord by a slender stalk containing blood vessels. The structures have been displaced to reveal the fetal surface of the placenta.

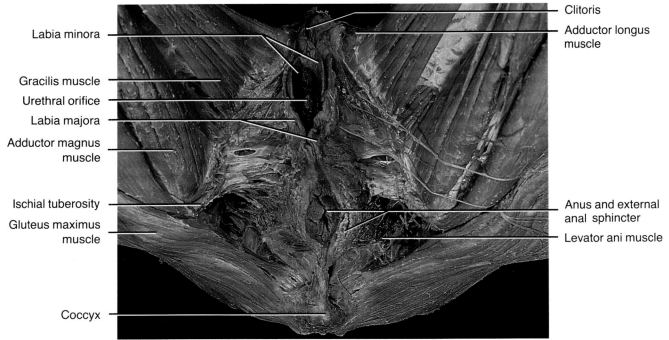

Labia minora

Gracilis muscle

Urethral orifice

Labia majora

Adductor magnus muscle

Ischial tuberosity

Gluteus maximus muscle

Coccyx

Clitoris

Adductor longus muscle

Anus and external anal sphincter

Levator ani muscle

Figure 5.8   Female perineum.

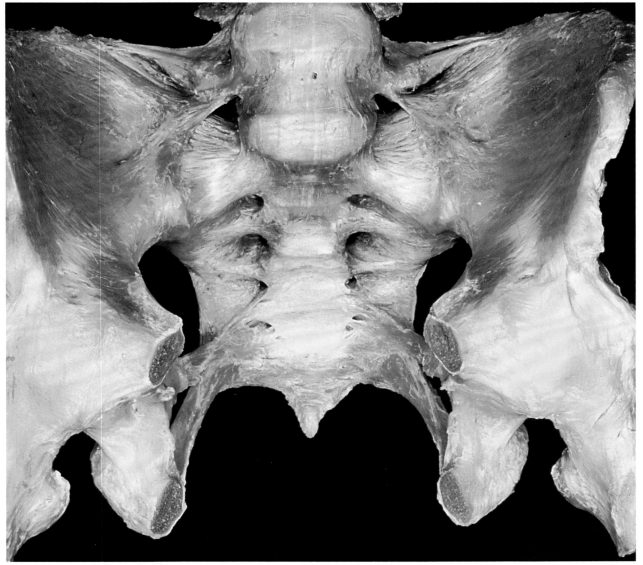

Figure 5.9A   Pelvic girdle ligaments.

Figure 5.8   Female perineum. A dissection of the female perineum shows the structures surrounding the anus and external genitalia. The superficial fascia has been removed from the perineum to show the deep perineal fascia, which appears as a distinct membranous layer over some of the important nerves to the external genitalia. On the left, the ischiorectal fossa has been dissected sufficiently to show the levator ani mucle deep to the external anal sphincter. The clitoris and labia minora are displayed in relationship to the external urethral orifice.

Figure 5.9A, B   Pelvic girdle ligaments. Ligaments for the pelvic girdle from the anterior view, with the pubic bones removed. Nonligamentous soft tissues have been removed from this specimen to show the tough ligamentous structures linking the pelvic bones to the axial skeleton, represented by the lower lumbar vertebra, the sacrum, and the coccyx. Iliolumbar ligaments superiorly and ligaments of the sacroiliac joints support the articulations between iliac bones and the sacrum. Inferiorly, the sacrotuberous and sacrospinal ligaments divide the sciatic notch into a greater and lesser sciatic foramen.

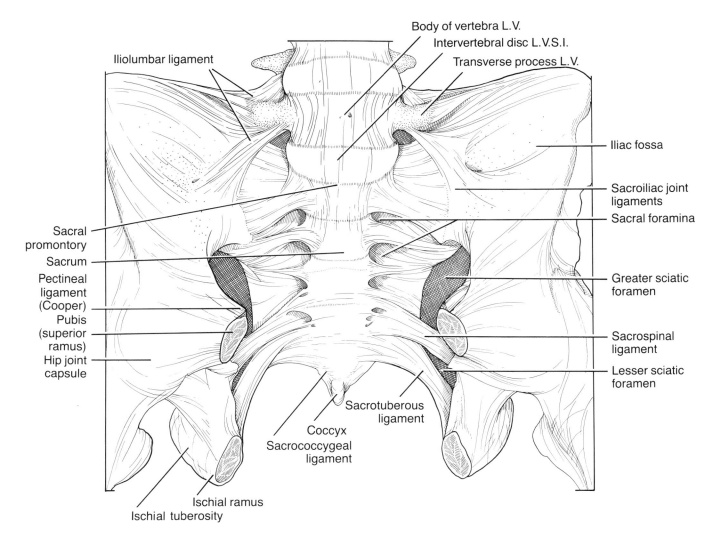

Figure 5.9B   Pelvic girdle ligaments.

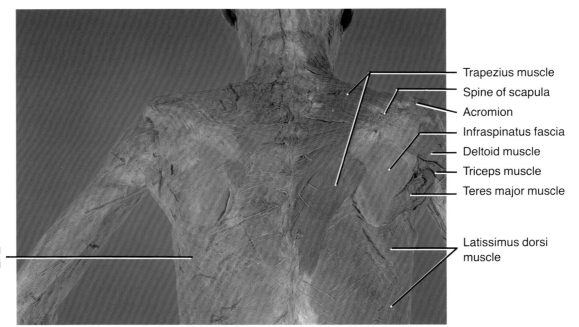

Trapezius muscle
Spine of scapula
Acromion
Infraspinatus fascia
Deltoid muscle
Triceps muscle
Teres major muscle

Latissimus dorsi muscle

Fascia covering latissimus dorsi

Figure 6.1   Shoulder girdle muscles of the back.

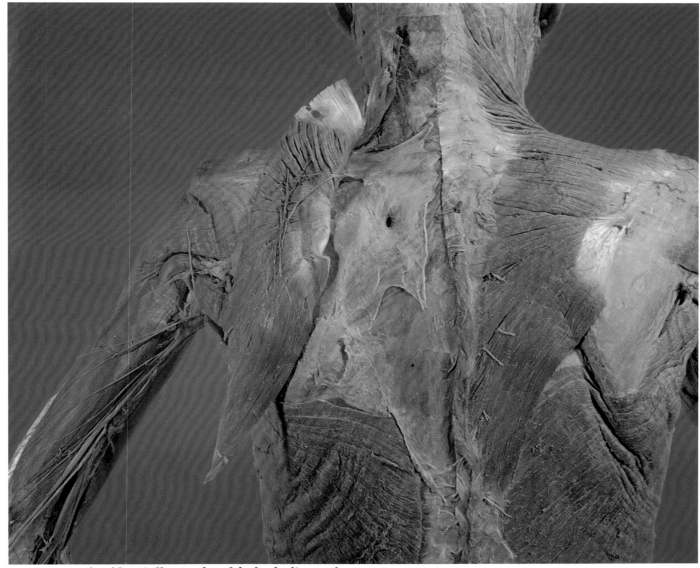

Figure 6.2A   Shoulder girdle muscles of the back, dissected.

Figure 6.1 Shoulder girdle muscles of the back. Dissection of the thoracic and lumbosacral regions of the back, preserving the deep fascia on the left and removing it on the right to expose the superficial layer of muscles. The multiple cutaneous nerves and vessels may be seen penetrating the deep fascia on the left. On the right, the trapezius muscle and the latissimus dorsi take their origin from the entire visible length of the fascia of the spinous processes of the cervical, thoracic, and lumbar vertebrae. The relationship of the teres major muscle to the latissimus dorsi and its origin from lateral border and lower angle of the scapula and insertion on the humerus are evident. The posterior portion of the deltoid muscle is visible. Beneath it, and extending over the entire dorsal aspect of the infraspinous portion of the scapula, is the infraspinatus muscle covered by infraspinatus fascia.

Figure 6.2A, B Shoulder girdle muscles of the back, dissected. This view of the superficial muscles of the back with the fascia removed displays an intact trapezius muscle on the right. The origins of the trapezius from the vertebral region have been sectioned and the muscle reflected laterally on the left to show its innervation by the accessory nerve and the fascia overlying the rhomboid muscles.

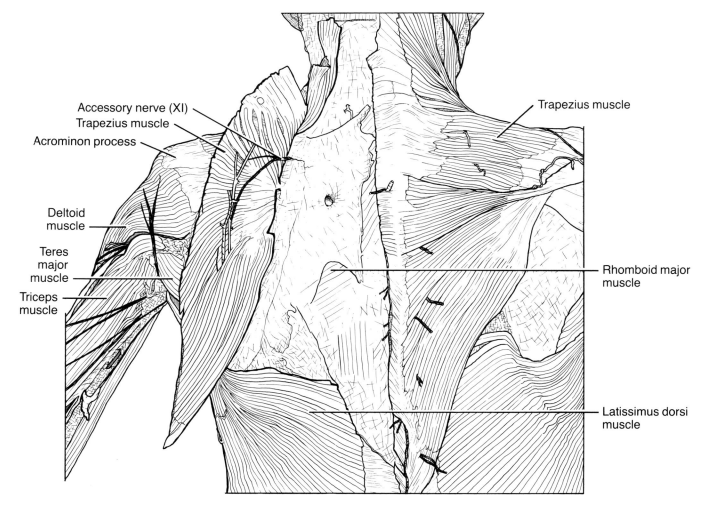

Figure 6.2B Shoulder girdle muscles of the back, dissected.

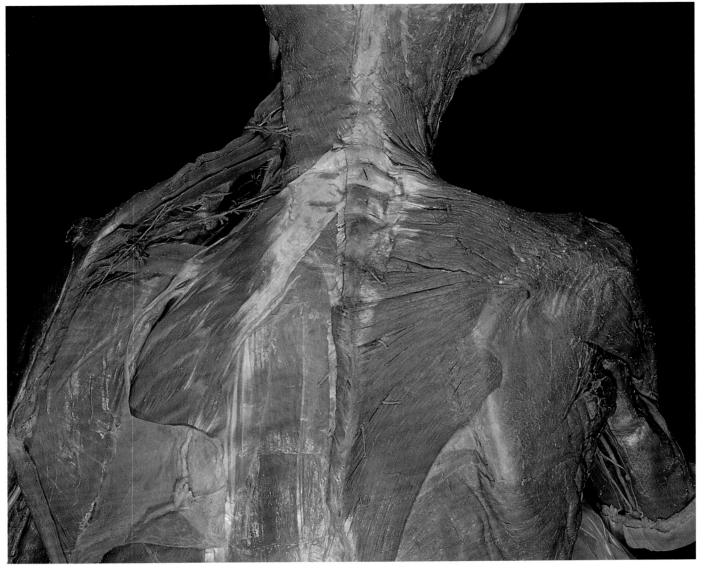

Figure 6.3A   Shoulder girdle muscles of the back, dissected.

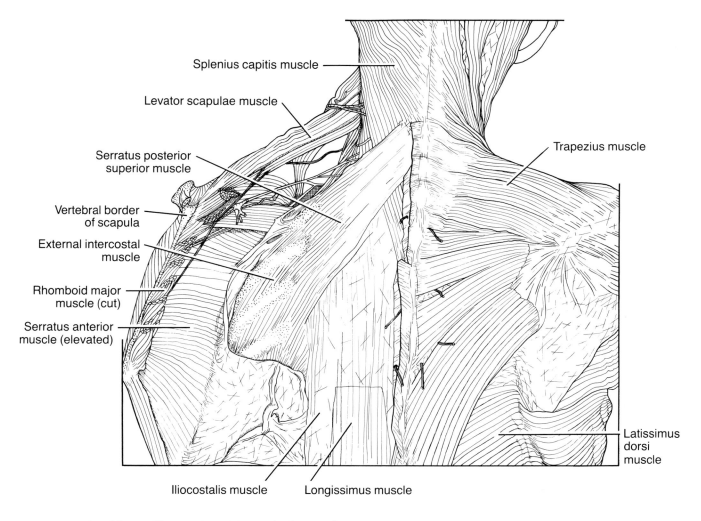

Splenius capitis muscle

Levator scapulae muscle

Serratus posterior
superior muscle

Vertebral border
of scapula

External intercostal
muscle

Rhomboid major
muscle (cut)

Serratus anterior
muscle (elevated)

Trapezius muscle

Latissimus
dorsi
muscle

Iliocostalis muscle

Longissimus muscle

Figure 6.3B   Shoulder girdle muscles of the back, dissected.

Figure 6.3A, B   Shoulder girdle muscles of the back, dissected. This dissection of the thoracic and lumbosacral regions of the back preserves the superficial muscular layer on the right while sectioning the trapezius and rhomboid muscles on the left, allowing elevation of the scapula from the chest wall. Beneath the scapula, one may see the serratus anterior muscle attaching along the vertebral border of the scapula and extending around the chest cage to attach as interdigitations along the outer surfaces of the upper eight or nine ribs and the intervening intercostal fascia. The serratus posterior superior muscle aids in elevation of the ribs in respiration. The levator scapulae muscle is shown, as are the cut edges of the rhomboid muscles.

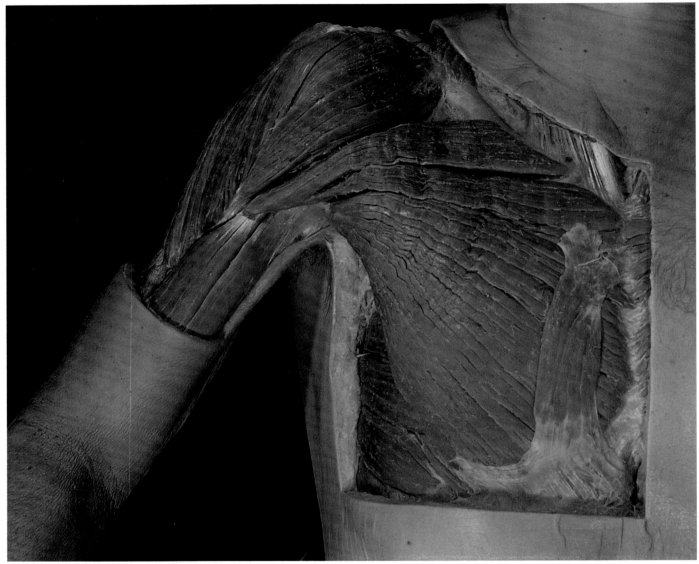

Figure 6.4A Pectoralis major and deltoid muscles.

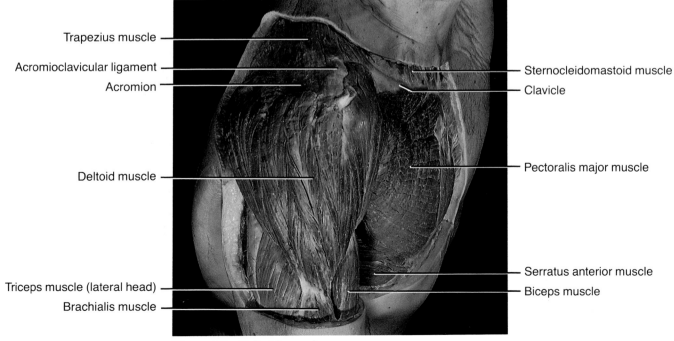

Trapezius muscle

Acromioclavicular ligament

Acromion

Deltoid muscle

Triceps muscle (lateral head)

Brachialis muscle

Sternocleidomastoid muscle

Clavicle

Pectoralis major muscle

Serratus anterior muscle

Biceps muscle

Figure 6.5 Deltoid muscle.

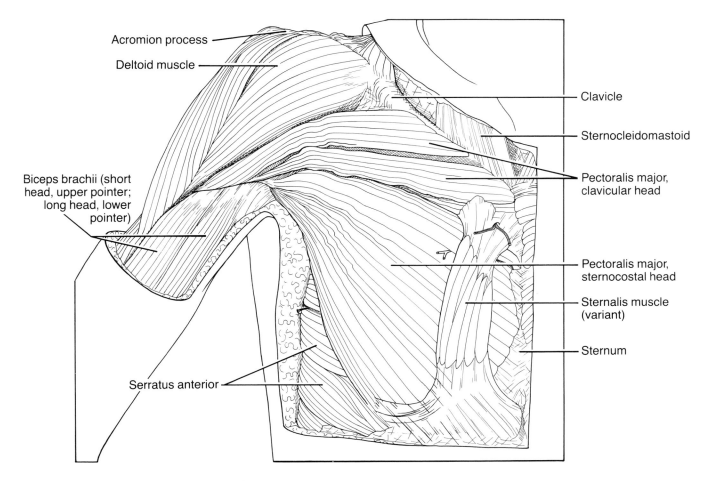

Acromion process

Deltoid muscle

Clavicle

Sternocleidomastoid

Pectoralis major, clavicular head

Biceps brachii (short head, upper pointer; long head, lower pointer)

Pectoralis major, sternocostal head

Sternalis muscle (variant)

Sternum

Serratus anterior

Figure 6.4B   Pectoralis major and deltoid muscles.

Figure 6.4A, B   Pectoralis major and deltoid muscles. This anterior view of the pectoralis major and deltoid muscles reveals the relationship of these structures to the clavicle, acromion process, sternum, and chest cage as well as to the muscles extending distally into the arm.

Figure 6.5   Deltoid muscle. This is a view of the right deltoid muscle with the skin and superficial and deep fascia removed from the deltoid and pectoral regions. The anterior and posterior parts of the deltoid muscle may be seen taking origin from the clavicle acromion process and spine of the scapula to insert as a common tendon on the lateral aspect of the humerus. The biceps in front, the triceps behind, and the brachialis muscle below the insertion of the deltoid are clearly shown. The clavicular and sternal origins of the pectoralis major muscle have been left intact.

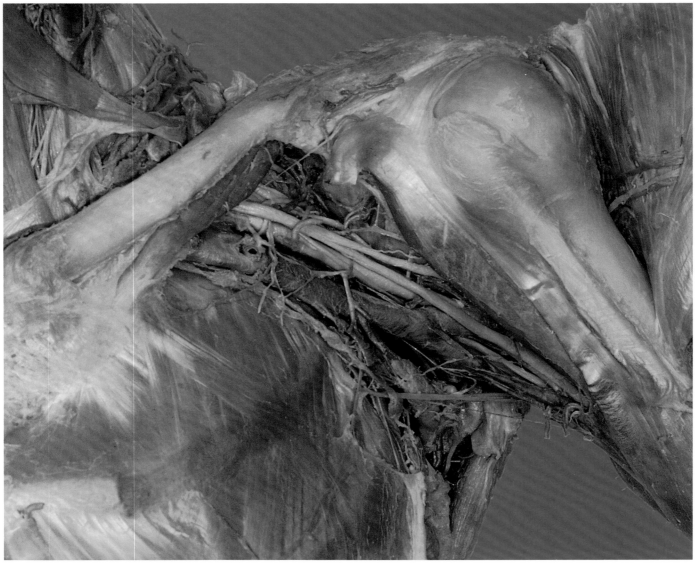

Figure 6.6A   Brachial plexus and axilla.

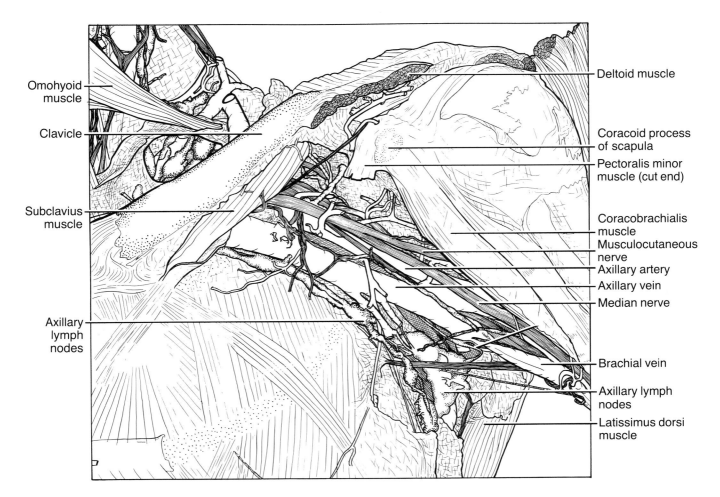

Omohyoid muscle

Clavicle

Subclavius muscle

Axillary lymph nodes

Deltoid muscle

Coracoid process of scapula

Pectoralis minor muscle (cut end)

Coracobrachialis muscle

Musculocutaneous nerve

Axillary artery

Axillary vein

Median nerve

Brachial vein

Axillary lymph nodes

Latissimus dorsi muscle

Figure 6.6B   Brachial plexus and axilla.

Figure 6.6A, B   Brachial plexus and axilla. This view shows the complex relationships of the numerous structures in the axilla lying beneath the pectoralis major and minor muscles, which have been resected. The subclavius muscle lying beneath the clavicle forms a protective cushion over the major vessels and nerves passing from the neck to the axilla. Notable in the axilla are the axillary vein, the axillary artery, components of the brachial plexus, and numerous axillary lymph nodes, some of which have been removed to display the axillary neurovascular structures.

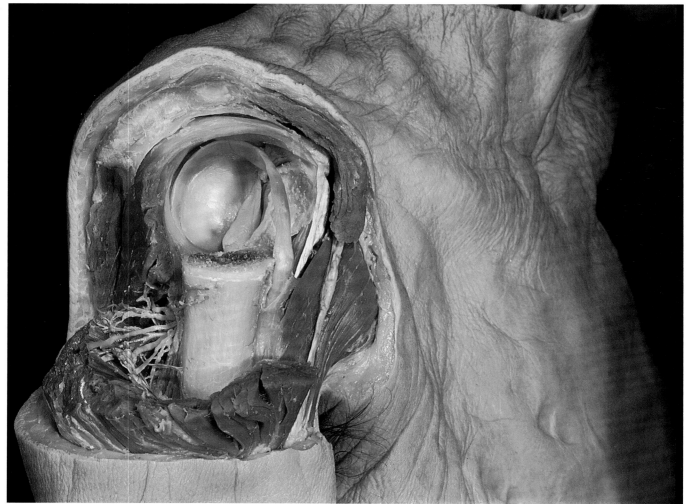

Figure 6.7A   Shoulder joint.

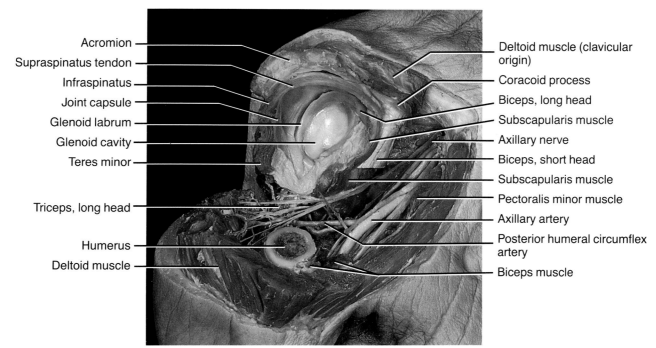

Acromion

Supraspinatus tendon

Infraspinatus

Joint capsule

Glenoid labrum

Glenoid cavity

Teres minor

Triceps, long head

Humerus

Deltoid muscle

Deltoid muscle (clavicular origin)

Coracoid process

Biceps, long head

Subscapularis muscle

Axillary nerve

Biceps, short head

Subscapularis muscle

Pectoralis minor muscle

Axillary artery

Posterior humeral circumflex artery

Biceps muscle

Figure 6.8   Shoulder joint.

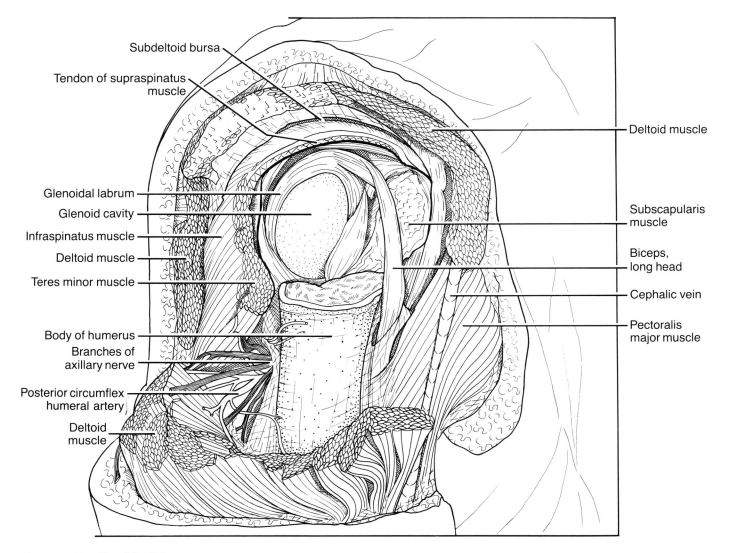

Subdeltoid bursa

Tendon of supraspinatus muscle

Glenoidal labrum

Glenoid cavity

Infraspinatus muscle

Deltoid muscle

Teres minor muscle

Body of humerus

Branches of axillary nerve

Posterior circumflex humeral artery

Deltoid muscle

Deltoid muscle

Subscapularis muscle

Biceps, long head

Cephalic vein

Pectoralis major muscle

Figure 6.7B   Shoulder joint.

Figure 6.7A, B   Shoulder joint. The cavity of the right shoulder joint is displayed by removal of the head of the humerus. Note the deepening of the glenoid cavity by the fibrocartilaginous glenoid labrum, which is itself linked to the long head of the biceps tendon. The cut edges of the teres minor, infraspinatus, supraspinatus, and subscapularis muscles form the rotator cuff group of muscles around the shoulder. The axillary nerve to the deltoid muscle, together with the posterior humeral circumflex artery, courses medial to the humerus at its neck.

Figure 6.8   Shoulder joint. This view of the shoulder is augmented by removal of the proximal humerus. It shows the relationships of structures passing behind the humerus and below the shoulder, including the posterior humeral circumflex artery and the axillary nerve. The axillary artery and brachial plexus nerves continue distally into the axilla.

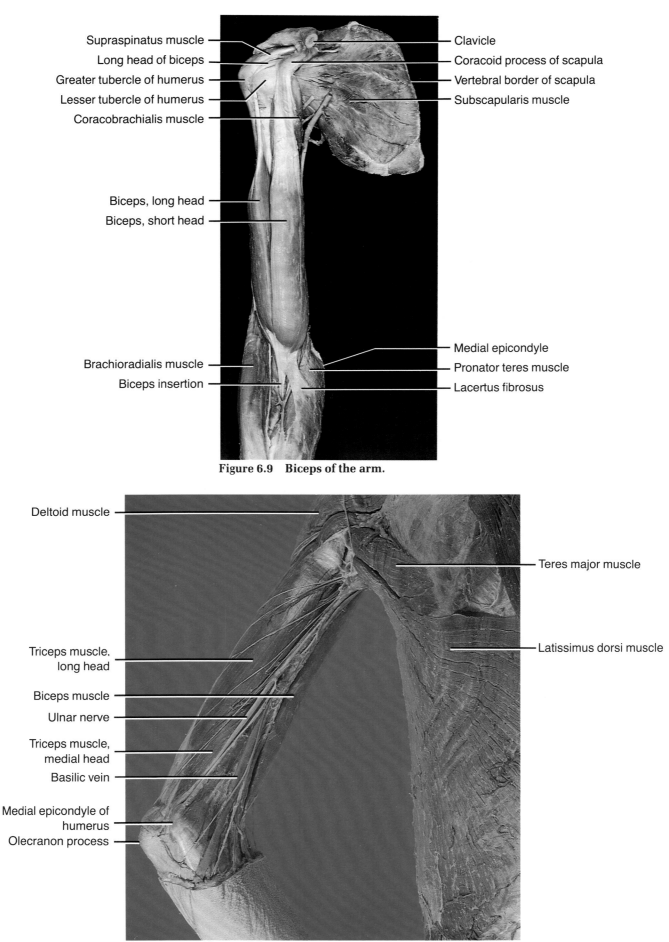

Supraspinatus muscle

Long head of biceps

Greater tubercle of humerus

Lesser tubercle of humerus

Coracobrachialis muscle

Clavicle

Coracoid process of scapula

Vertebral border of scapula

Subscapularis muscle

Biceps, long head

Biceps, short head

Brachioradialis muscle

Biceps insertion

Medial epicondyle

Pronator teres muscle

Lacertus fibrosus

Figure 6.9   Biceps of the arm.

Deltoid muscle

Teres major muscle

Triceps muscle, long head

Biceps muscle

Ulnar nerve

Triceps muscle, medial head

Basilic vein

Medial epicondyle of humerus

Olecranon process

Latissimus dorsi muscle

Figure 6.10   Triceps of the arm.

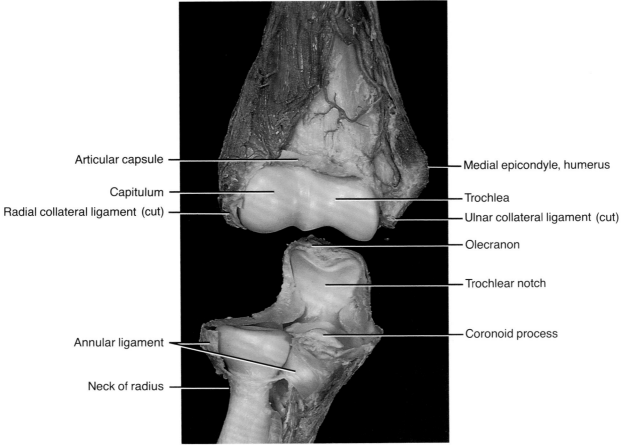

Articular capsule

Capitulum

Radial collateral ligament (cut)

Annular ligament

Neck of radius

Medial epicondyle, humerus

Trochlea

Ulnar collateral ligament (cut)

Olecranon

Trochlear notch

Coronoid process

Figure 6.11   Elbow joint.

Figure 6.9   Biceps of the arm. This anterior view shows the relationships of the important biceps muscle. Note the long head passing to the supraglenoid tubercle and the short head attaching to the corticoid process. The distal sites of insertion are into the deep fascia of the forearm, through the lacertus fibrosus, and by the tendinous insertion into the biceps tuberosity of the radius.

Figure 6.10   Triceps of the arm. Looking at the left triceps muscle from the posteromedial view, one sees its long head coursing upward toward its insertion on the inferior glenoid tubercle. The medial and lateral heads take origin from the humerus. The three heads of the triceps coalesce into the triceps tendon to attach to the olecranon process of the ulna. Note the protected position of the major nerve and vascular structures on the medial aspect of the arm along the intermuscular septum between the triceps and biceps muscles.

Figure 6.11   Elbow joint. The articular cavity of the elbow joint has been opened to view it from the anterior side. Note the complex set of articulations between the humerus, the radius, and the ulna. The trochlear notch of the ulna articulates with the trochlea of the distal humerus, and the radial head articulates with the capitulum. The radial head also articulates with the ulna to allow rotation of the radius across the ulna in pronation and a return in supination. The articular head of the radius is held in place by the tough annular ligament. The radial and ulnar collateral ligaments have been sectioned to disarticulate the joint.

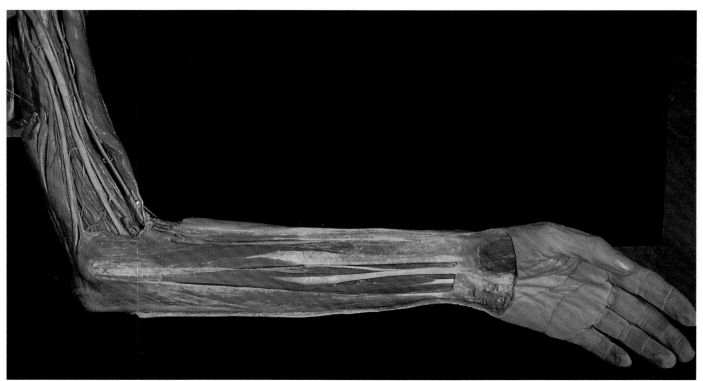

Figure 6.12A   Palmar forearm.

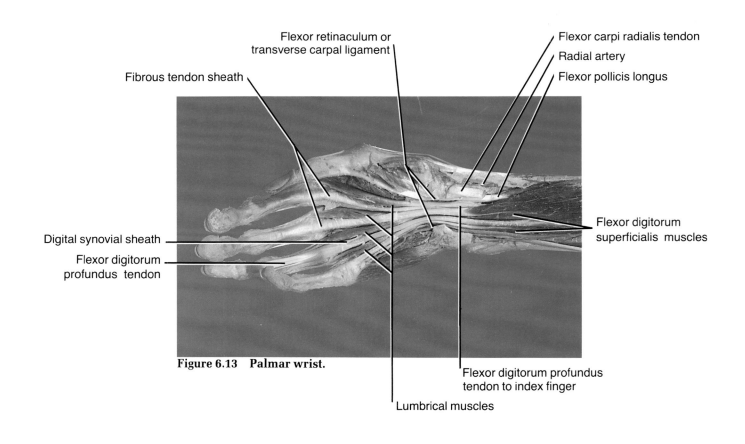

Flexor retinaculum or
transverse carpal ligament

Flexor carpi radialis tendon

Radial artery

Flexor pollicis longus

Fibrous tendon sheath

Digital synovial sheath

Flexor digitorum
profundus tendon

Flexor digitorum
superficialis muscles

Figure 6.13   Palmar wrist.

Flexor digitorum profundus
tendon to index finger

Lumbrical muscles

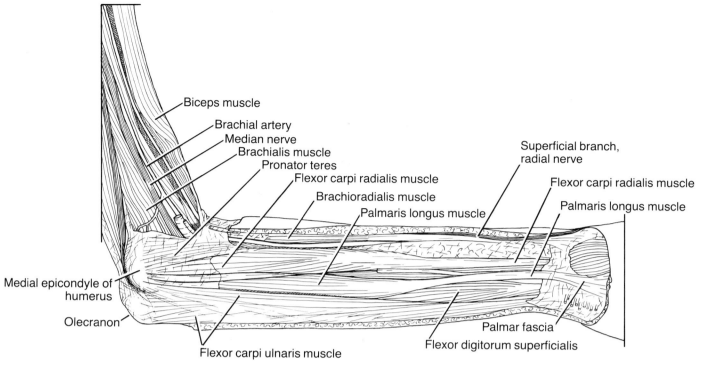

Biceps muscle
Brachial artery
Median nerve
Brachialis muscle
Pronator teres
Flexor carpi radialis muscle
Brachioradialis muscle
Palmaris longus muscle

Superficial branch, radial nerve
Flexor carpi radialis muscle
Palmaris longus muscle

Medial epicondyle of humerus

Olecranon

Flexor carpi ulnaris muscle

Palmar fascia
Flexor digitorum superficialis

Figure 6.12B Palmar forearm.

Figure 6.12A, B   Palmar forearm. The volar aspect of the forearm shows the superficial group of muscles consisting of the pronator teres, flexor carpi radialis, palmaris longus, and flexor carpi ulnaris. The middle layer seen just beneath the palmaris longus consists of the fleshy group of flexor digitorum superficialis muscles. The deep layer consisting of flexor digitorum profundi, flexor pollicus longus, and pronator quadratus cannot be seen. Note the palmar fascia taking its origin from the palmaris longus tendon.

Figure 6.13   Palmar wrist. The palmar aspect of the right hand with the carpal tunnel open shows the structures that pass beneath the transverse carpal ligament at the wrist. The median nerve and blood vessels have been removed, except for a part of the ulnar artery and the radial artery above the wrist. The most superficial structures remaining in the carpal tunnel are the flexor digitorum superficialis tendons. The deep layer, consisting of the flexor pollicis longus and the flexor digitorum profundus tendons, are represented here by the flexor pollicis longus and the flexor digitorum profundus tendon to the index finger. The lumbrical muscles are seen taking origin from the profundus tendons in the palm.

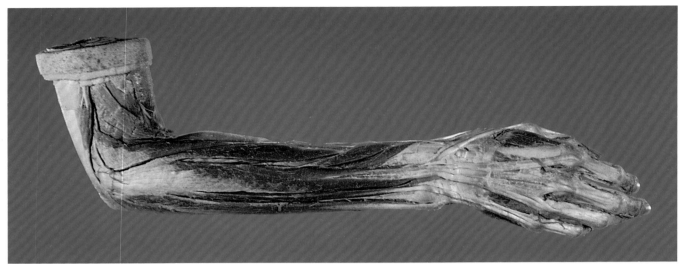

Figure 6.14A   Dorsal forearm.

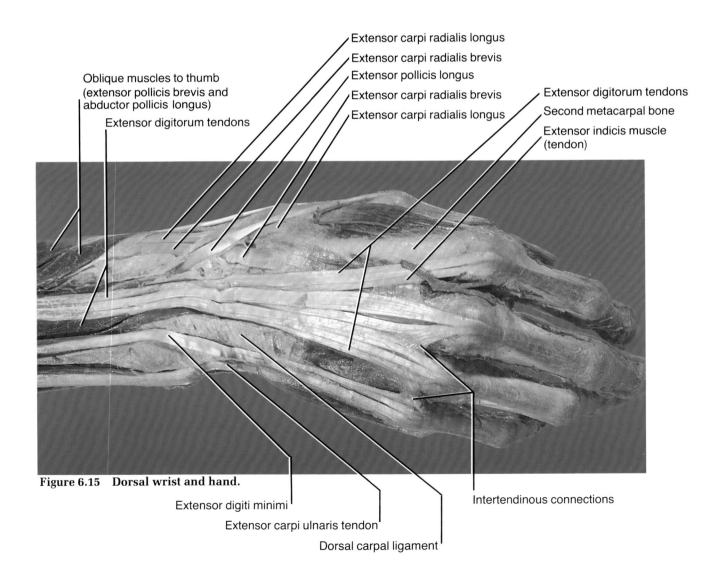

Oblique muscles to thumb
(extensor pollicis brevis and
abductor pollicis longus)

Extensor digitorum tendons

Extensor carpi radialis longus

Extensor carpi radialis brevis

Extensor pollicis longus

Extensor carpi radialis brevis

Extensor carpi radialis longus

Extensor digitorum tendons

Second metacarpal bone

Extensor indicis muscle
(tendon)

Figure 6.15   Dorsal wrist and hand.

Extensor digiti minimi

Extensor carpi ulnaris tendon

Dorsal carpal ligament

Intertendinous connections

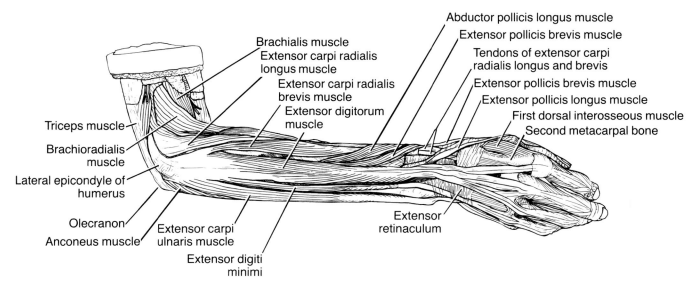

Brachialis muscle
Extensor carpi radialis longus muscle
Extensor carpi radialis brevis muscle
Extensor digitorum muscle
Triceps muscle
Brachioradialis muscle
Lateral epicondyle of humerus
Olecranon
Anconeus muscle
Extensor carpi ulnaris muscle
Extensor digiti minimi
Abductor pollicis longus muscle
Extensor pollicis brevis muscle
Tendons of extensor carpi radialis longus and brevis
Extensor pollicis brevis muscle
Extensor pollicis longus muscle
First dorsal interosseous muscle
Second metacarpal bone
Extensor retinaculum

Figure 6.14B   Dorsal forearm.

Figure 6.14A, B   Dorsal forearm. The extensor or dorsal aspect of the forearm displays the long extensor muscles to the fingers and thumb, the major wrist extensor muscles, and the outcropping oblique muscles to the thumb (the abductor pollicis longus and extensor pollicis brevis muscles). The brachioradialis muscle is seen crossing the elbow, where it acts with the brachialis and biceps to flex the elbow. These are antagonistic to the triceps muscle, the major extensor of the elbow. Note that the extensor carpi ulnaris muscle is positioned to act both as wrist extensor and ulnar deviator.

Figure 6.15   Dorsal wrist and hand. Extensor tendons at the wrist pass through six separate compartments beneath the extensor retinaculum, called the dorsal carpal ligament. Each compartment contains tendons that represent either one or two muscles. The number of muscles in each of the six compartments is listed below, starting from the radial side of the wrist.

Number of Muscles	Compartment	Muscles
2	First	Abductor pollicis longus and extensor pollicis brevis
2	Second	Extensor carpi radialis longus and brevis
1	Third	Extensor pollicis longus
2	Fourth	Extensor digitorum and extensor indicis
1	Fifth	Extensor digit minimi
1	Sixth	Extensor carpi ulnaris

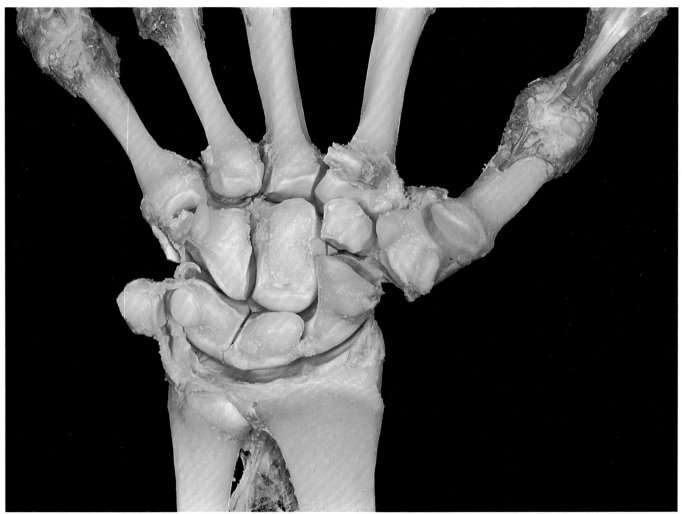

Figure 6.16A   Wrist bones.

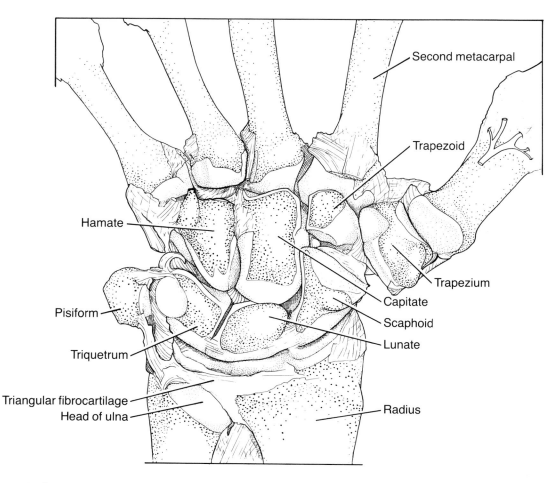

Figure 6.16B   Wrist bones.

Figure 6.16A, B   Wrist bones. This palmar or volar view of the right wrist with the carpal bones separated displays the eight carpal bones and their relationship to one another and to the radius and ulna proximally and to the metacarpals distally. Note that the articulation between forearm bones and carpals is between carpals and the expanded distal radius and with the head of the ulna through the triangular fibrocartilage.

7 The Lower Limb

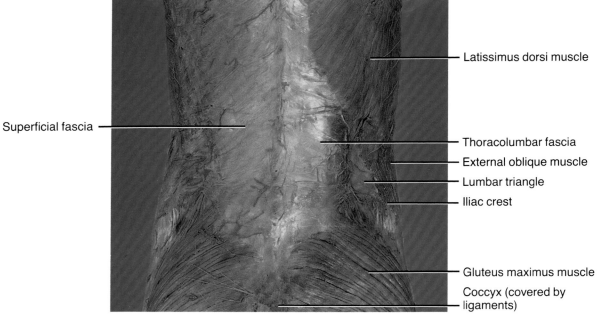

Latissimus dorsi muscle

Superficial fascia

Thoracolumbar fascia

External oblique muscle

Lumbar triangle

Iliac crest

Gluteus maximus muscle

Coccyx (covered by ligaments)

Figure 7.1 Back and upper gluteal region.

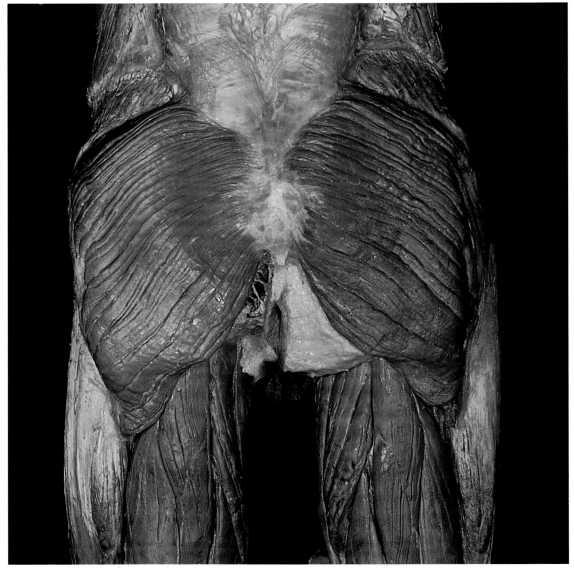

Figure 7.2A Gluteal region.

Figure 7.1 **Back and upper gluteal region.** The dissection of the thoracic and lumbosacral regions of the back preserves the superficial fascia on the left with multiple cutaneous nerves and vessels. On the right, the origin of the latissimus dorsi from the thoracolumbar fascia is readily seen. The fascia covers back muscles that, together with abdominal wall muscles, control relationships of the pelvis and axial skeleton.

Figure 7.2A, B **Gluteal region.** The fascia covering the gluteus maximus muscles has been removed, together with superficial vessels and nerves of the buttocks. The fat has been removed from within the left ischiorectal fossa, and the posterior part of the fascia lata has been cut away to expose the posterior musculature of the leg.

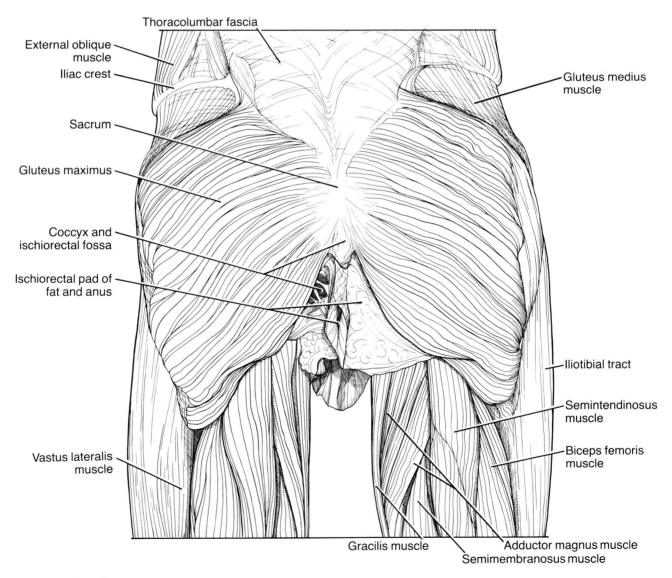

Thoracolumbar fascia

External oblique muscle

Iliac crest

Sacrum

Gluteus maximus

Coccyx and ischiorectal fossa

Ischiorectal pad of fat and anus

Vastus lateralis muscle

Gluteus medius muscle

Iliotibial tract

Semintendinosus muscle

Biceps femoris muscle

Gracilis muscle

Adductor magnus muscle

Semimembranosus muscle

Figure 7.2B **Gluteal region.**

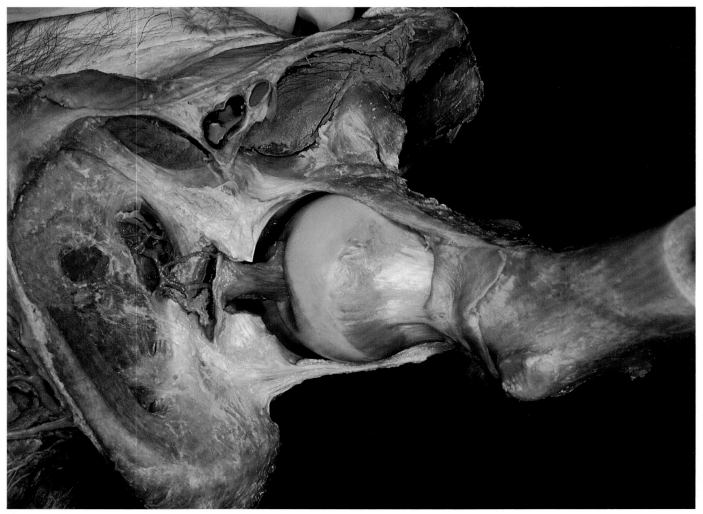

Figure 7.3A   Hip joint.

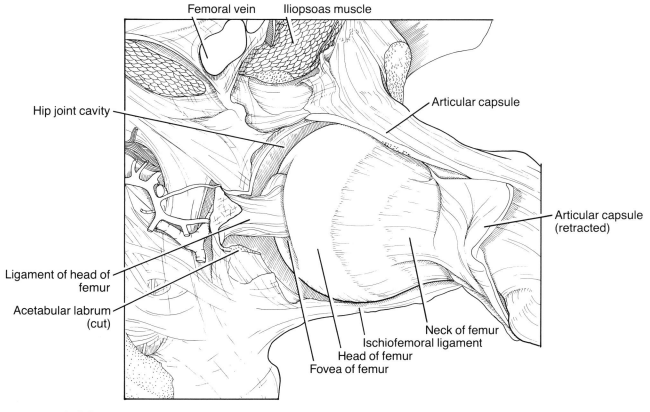

Femoral vein   Iliopsoas muscle

Articular capsule

Hip joint cavity

Articular capsule
(retracted)

Ligament of head of
femur

Acetabular labrum
(cut)

Neck of femur

Ischiofemoral ligament

Head of femur

Fovea of femur

Figure 7.3B   Hip joint.

Figure 7.3A, B   Hip joint. A large window has been cut through the inferior part of the hip joint capsule to expose the synovial cavity of the joint. The capsule is retracted laterally, exposing the ligamentous portions of the articular capsule both superiorly and inferiorly. The depression in the head of the femur, called the fovea, is the site of attachment of the large ligament of the head of the femur, which carries an acetabular branch of the obturator artery. The cut edge of the iliopsoas muscle and femoral vessels may be seen passing beneath the inguinal ligament above the hip joint. The socket of the joint is deepened by the acetabular labrum, which has been sectioned to show the ligament of the head of the femur more clearly.

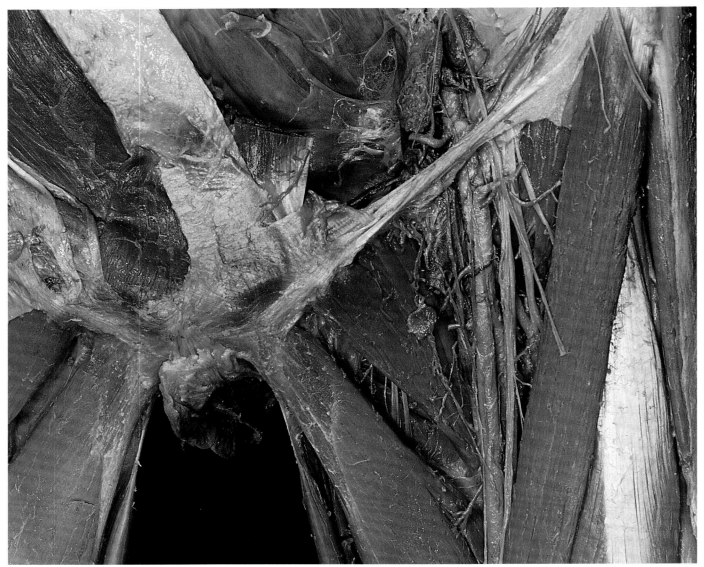

Figure 7.4A Inguinal area.

External iliac artery

Inguinal ligament

Femoral artery

Femoral vein

Pectineus muscle

Superficial femoral artery

Sartorius muscle (reflected)

Adductor longus muscle

Sartorius muscle (origin)

Femoral nerve

Iliacus muscle

Tensor fascia lata muscle

Profunda femoris artery

Rectus femoris muscle

Vastus intermedius muscle

Figure 7.5 Inguinal area.

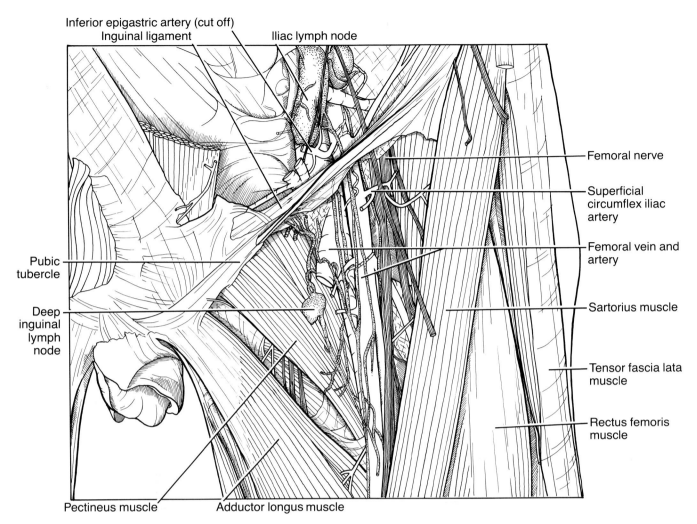

Inferior epigastric artery (cut off)
Inguinal ligament
Iliac lymph node
Femoral nerve
Superficial circumflex iliac artery
Femoral vein and artery
Pubic tubercle
Sartorius muscle
Deep inguinal lymph node
Tensor fascia lata muscle
Rectus femoris muscle
Pectineus muscle
Adductor longus muscle

Figure 7.4B   Inguinal area.

Figure 7.4A, B   Inguinal area. Dissection of the femoral inguinal area displays the contents of the femoral triangle and structures passing beneath the inguinal ligament. The femoral triangle is formed by the inguinal ligament above, the sartorius muscle laterally, and the adductor longus muscle medially. Femoral vessels and femoral nerves enter the triangle beneath the inguinal ligament enclosed in the femoral sheath (dissected). Inguinal lymph nodes and lymphatic vessels are prominent in the femoral triangle.

Figure 7.5   Inguinal area. Dissection of the anterior thigh reveals structures passing beneath the inguinal ligament. Lateral to the femoral vein and femoral artery, the femoral nerve passes into the thigh, where it arborizes into multiple muscular branches to all the hip flexor, knee extensor group of muscles. The sartorius muscle has been reflected medially from its normal course obliquely from the anterior superior iliac spine to the medial aspect of the knee. This exposes the subsartorial structures, including the superficial femoral vessels.

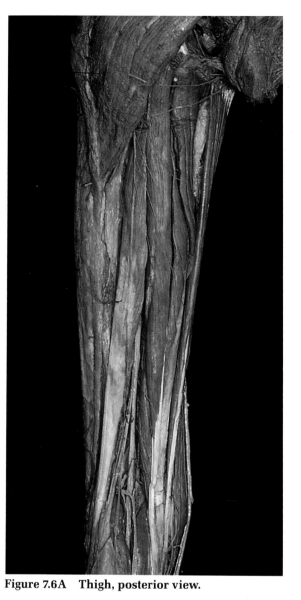

Figure 7.6A   Thigh, posterior view.

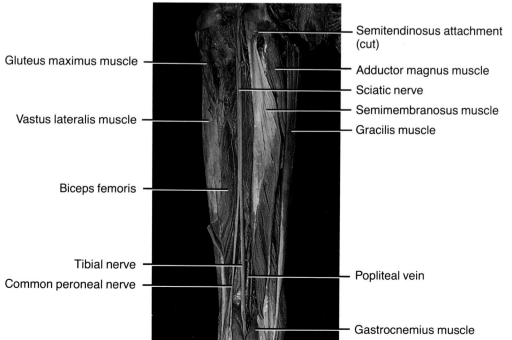

Gluteus maximus muscle —

Vastus lateralis muscle —

Biceps femoris —

Tibial nerve —
Common peroneal nerve —

— Semitendinosus attachment (cut)

— Adductor magnus muscle
— Sciatic nerve
— Semimembranosus muscle
— Gracilis muscle

— Popliteal vein

— Gastrocnemius muscle

Figure 7.7   Thigh, posterior view.

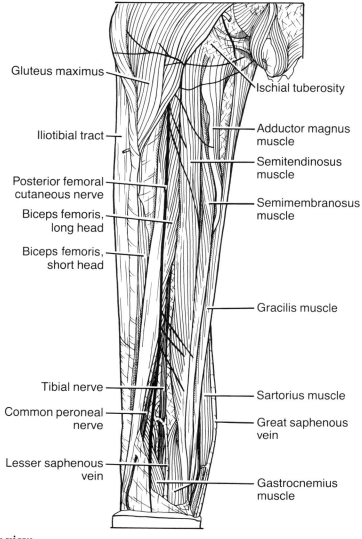

Gluteus maximus

Iliotibial tract

Posterior femoral
cutaneous nerve

Biceps femoris,
long head

Biceps femoris,
short head

Tibial nerve

Common peroneal
nerve

Lesser saphenous
vein

Ischial tuberosity

Adductor magnus
muscle

Semitendinosus
muscle

Semimembranosus
muscle

Gracilis muscle

Sartorius muscle

Great saphenous
vein

Gastrocnemius
muscle

Figure 7.6B   Thigh, posterior view.

Figure 7.6A, B   Thigh, posterior view. This dissection of the posterior aspect of the left thigh shows the intact hamstring muscles. The fascia lata has been removed to expose the muscles that make up the hamstring group, including the biceps femoris, semitendinosus, and semimembranosus.

Figure 7.7   Thigh, posterior view. This dissection of the posterior aspect of the left leg displays the hamstring muscles and their relationships. The gluteus maximus muscle has been removed to expose the proximal attachments of the semitendinosus, which has been removed close to its attachment at the ischial tuberosity, and the semimembranosus, which remains intact. Laterally, the biceps femoris may be seen. Between the medial lateral hamstring muscles, the sciatic nerve descends, splitting into its tibial and common peroneal nerve branches.

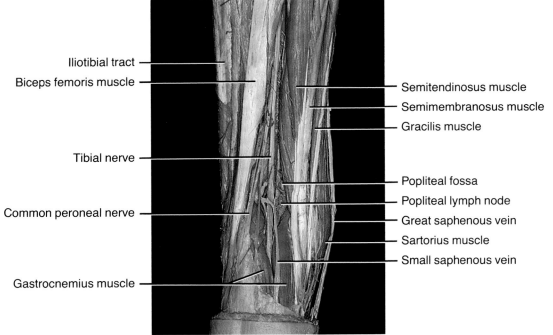

Iliotibial tract —

Biceps femoris muscle —

Tibial nerve —

Common peroneal nerve —

Gastrocnemius muscle —

— Semitendinosus muscle

— Semimembranosus muscle

— Gracilis muscle

— Popliteal fossa

— Popliteal lymph node

— Great saphenous vein

— Sartorius muscle

— Small saphenous vein

Figure 7.8   Popliteal fossa.

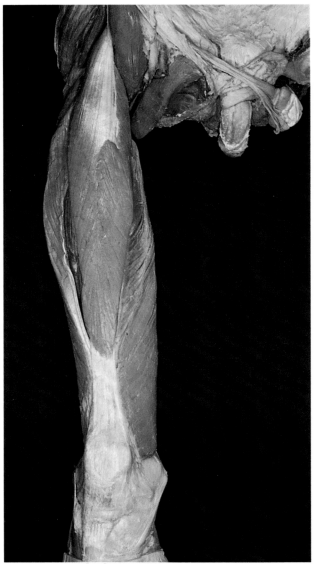

Figure 7.9A   Thigh, anterior view.

Figure 7.8 Popliteal fossa. This dissection of the posterior aspect of the left thigh and popliteal fossa demonstrates the hamstring muscles and their relationship to the sciatic nerve and its branches, as well as the origins of the gastrocnemius muscle. Fat has been removed from the popliteal fossa, preserving the small saphenous vein and lymph nodes.

Figure 7.9A, B Thigh, anterior view. The muscles of the anterior and medial aspects of the thigh include the quadriceps femoris muscle with its components. Their relationships to one another and insertion into the patella are displayed here.

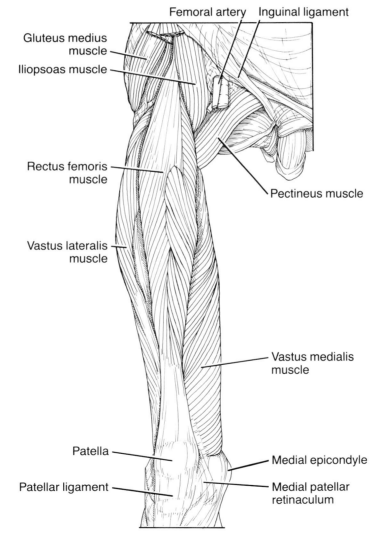

Femoral artery   Inguinal ligament

Gluteus medius muscle

Iliopsoas muscle

Rectus femoris muscle

Pectineus muscle

Vastus lateralis muscle

Vastus medialis muscle

Patella

Medial epicondyle

Patellar ligament

Medial patellar retinaculum

Figure 7.9B   Thigh, anterior view.

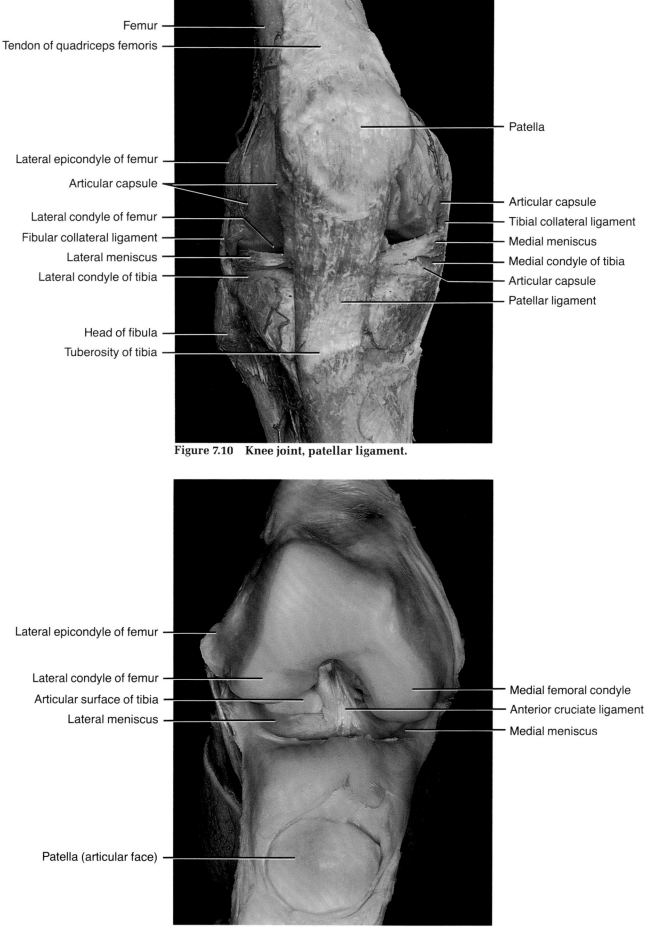

Femur

Tendon of quadriceps femoris

Patella

Lateral epicondyle of femur

Articular capsule

Articular capsule

Tibial collateral ligament

Lateral condyle of femur

Fibular collateral ligament

Medial meniscus

Lateral meniscus

Medial condyle of tibia

Lateral condyle of tibia

Articular capsule

Patellar ligament

Head of fibula

Tuberosity of tibia

Figure 7.10   Knee joint, patellar ligament.

Lateral epicondyle of femur

Lateral condyle of femur

Articular surface of tibia

Medial femoral condyle

Lateral meniscus

Anterior cruciate ligament

Medial meniscus

Patella (articular face)

Figure 7.11   Knee joint, open.

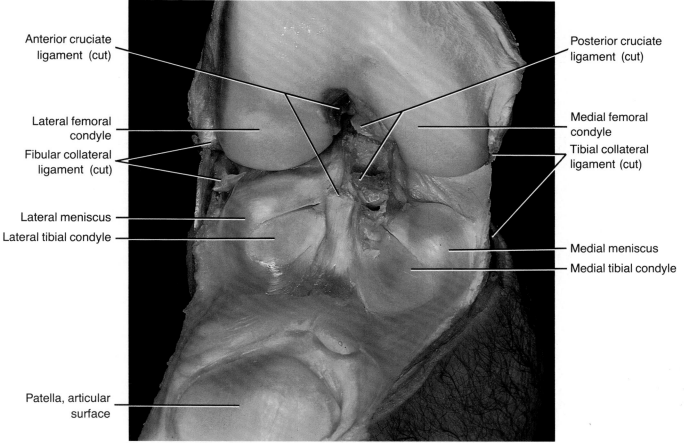

Anterior cruciate ligament (cut)

Posterior cruciate ligament (cut)

Lateral femoral condyle

Medial femoral condyle

Fibular collateral ligament (cut)

Tibial collateral ligament (cut)

Lateral meniscus

Lateral tibial condyle

Medial meniscus

Medial tibial condyle

Patella, articular surface

Figure 7.12   Knee joint, cruciates cut.

Figure 7.10   Knee joint, patellar ligament. The knee is partially dissected to show relationships of the patella and its ligaments in situ to the structures visible in the anterior portion of the joint cavity. The lateral fibular collateral ligament and the medial tibial collateral ligament lend important stability to the joint. The suprapatellar tendon represents the coalition of the muscles making up the quadriceps femoris. The patellar ligament below this sesamoid bone attaches to the tuberosity of the tibia, through which its muscles act as a knee extensor. The cut edges of the lateral meniscus and medial meniscus are visible.

Figure 7.11   Knee joint, open. This interior view of the right knee joint with the patella and its tendon reflected downward shows the interior of the right knee joint. The anterior cruciate ligament is prominent, and one may see the anterior extreme ends of the medial and lateral menisci.

Figure 7.12   Knee joint, cruciates cut. This view of the interior of the right knee joint, with the cruciate ligaments divided to allow separation of the femur from the tibia, displays the lateral and medial menisci along their entire superior surfaces. They serve to deepen the articular concavity on the superior surfaces of the tibia to receive the medial and lateral femoral condyles. The fibular collateral ligament laterally and the tibial collateral ligament medially have been sectioned to allow separation of femur and tibia to visualize the articular surfaces of the knee joint.

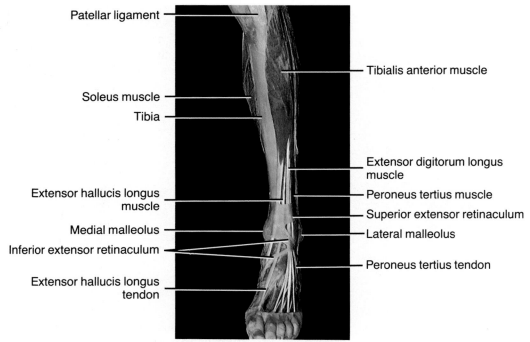

Patellar ligament

Tibialis anterior muscle

Soleus muscle

Tibia

Extensor digitorum longus muscle

Extensor hallucis longus muscle

Peroneus tertius muscle

Superior extensor retinaculum

Medial malleolus

Lateral malleolus

Inferior extensor retinaculum

Peroneus tertius tendon

Extensor hallucis longus tendon

Figure 7.13   Leg, anterior view.

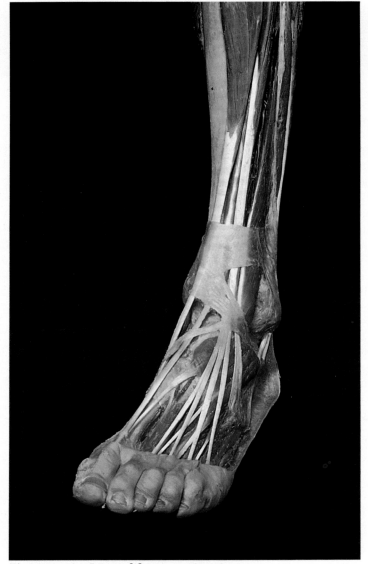

Figure 7.14A   Leg and foot, anterior view.

Figure 7.13 Leg, anterior view. Dissection of the anterior aspect of the left leg reveals the muscles of the leg and foot as seen from the front. The superficial and deep investing fascia, vessels, and nerves have been removed, with the exception of the thickened extensor retinacula, which have been retained across the ankle area. The subcutaneous medial surface of the tibia separates the structures in the anterior and posterior compartments in this view. The extensive superior and inferior extensor retinacula keep the extensor tendons from bowstringing across the angle between the leg and foot.

Figure 7.14A, B Leg and foot, anterior view. An anterolateral view of the leg and foot reveals muscles in the anterior, or extensor, compartment of the leg as they pass into the foot beneath the superior and inferior extensor retinacula. The intrinsic extensor digitorum brevis muscle, with its multiple tendons, and the extensor hallucis brevis muscle lie deep to the long extensors to the toes. The peroneus longus and brevis tendons are seen emerging from behind the lateral malleolus.

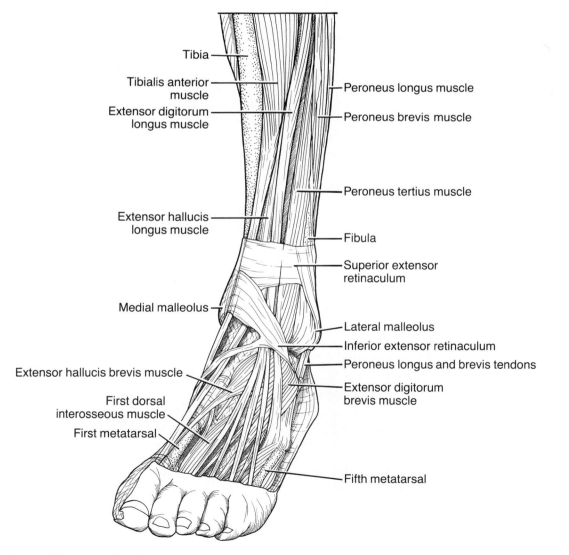

Figure 7.14B Leg and foot, anterior view.

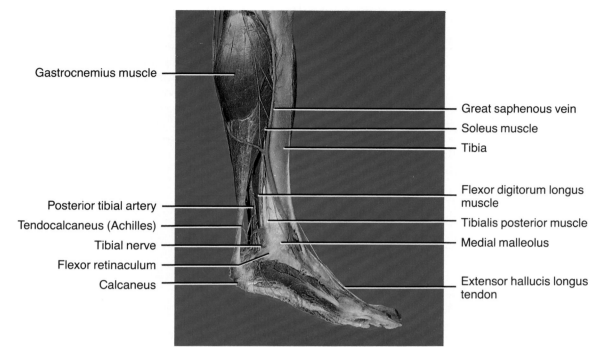

Gastrocnemius muscle

Great saphenous vein

Soleus muscle

Tibia

Flexor digitorum longus muscle

Posterior tibial artery

Tibialis posterior muscle

Tendocalcaneus (Achilles)

Tibial nerve

Medial malleolus

Flexor retinaculum

Calcaneus

Extensor hallucis longus tendon

Figure 7.15   Leg, medial view.

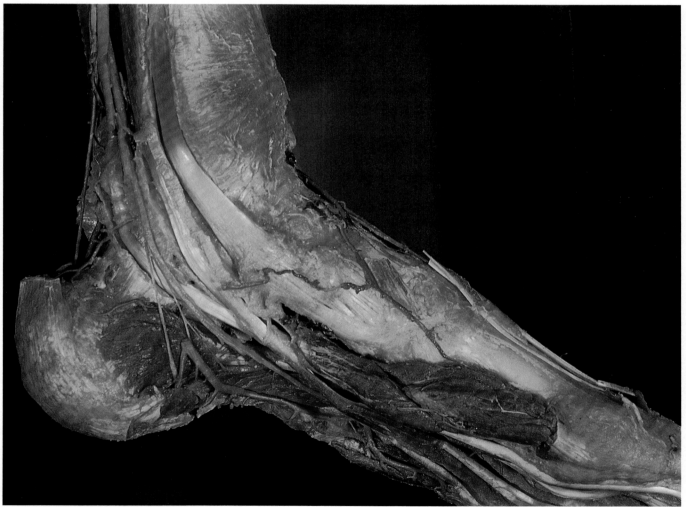

Figure 7.16A   Foot, medial view.

Figure 7.15 Leg, medial view. This dissection of the medial aspect of the left leg shows removal of the deep fascia of the leg and foot but preserves the large saphenous vein and superficial nerves to show their relationships to deeper structures. The great saphenous vein originating in the dorsum of the foot passes anterior to the medial malleolus and upward just behind the subcutaneous medial surface of the tibia toward the medial aspect of the knee. This view shows the relationship of the gastrocnemius and soleus, both of which insert into the calcaneus via the tendocalcaneus. Deep to these structures are the flexor digitorum longus, tibialis posterior, and the flexor hallucis longus, which is not visible in this view. Note the posterior tibial artery and veins and the tibial nerve as they pass behind the medial malleolus beneath the flexor retinaculum in the tarsal tunnel.

Figure 7.16A, B Foot, medial view. This dissection of the medial side of the ankle region displays the structures passing beneath the flexor retinaculum to the foot. Passing behind the medial malleolus are the tibialis posterior tendon, the tendons of the flexor digitorum longus and flexor hallucis longus, the tibial nerve, and the posterior tibial artery on its way to becoming the plantar arteries. The area roofed by the flexor retinaculum is commonly called the tarsal tunnel.

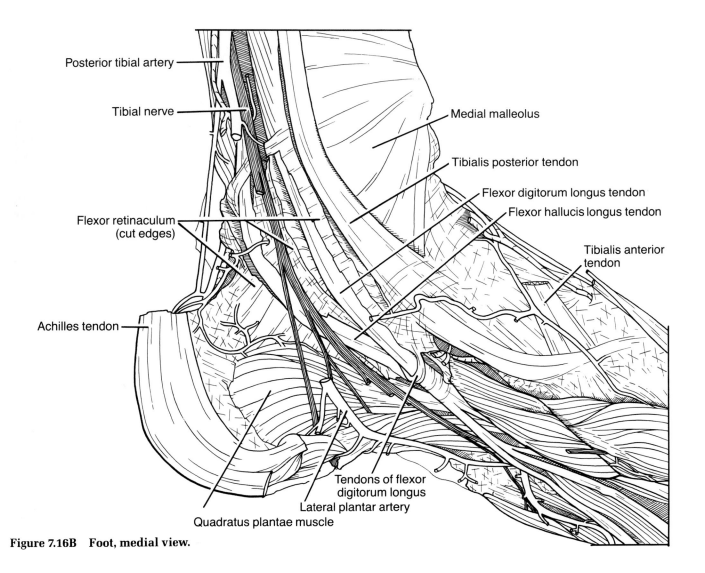

Figure 7.16B Foot, medial view.

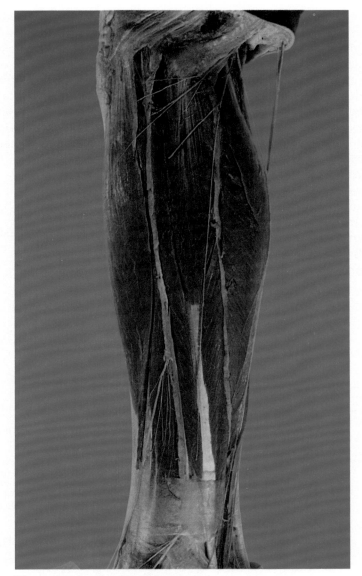

Figure 7.17A   Leg, lateral view.

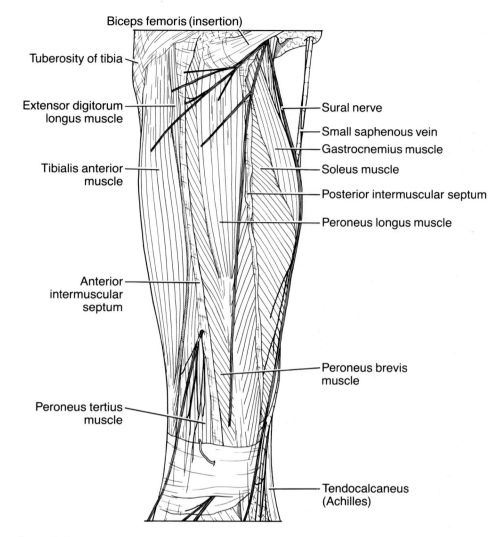

Biceps femoris (insertion)

Tuberosity of tibia

Extensor digitorum
longus muscle

Tibialis anterior
muscle

Anterior
intermuscular
septum

Peroneus tertius
muscle

Sural nerve

Small saphenous vein

Gastrocnemius muscle

Soleus muscle

Posterior intermuscular septum

Peroneus longus muscle

Peroneus brevis
muscle

Tendocalcaneus
(Achilles)

Figure 7.17B   Leg, lateral view.

Figure 7.17A, B   Leg, lateral view. This dissection of the lateral aspect of the left leg, after removal of the superficial vessels and fascia, reveals the two lateral intermuscular septa, one anteriorly between the anterior and lateral muscle groups and the second posteriorly between the lateral and posterior muscle groups. Visible here in the anterior group are the extensor digitorum longus, tibialis anterior, and peroneus tertius muscles. The peroneus longus and peroneus brevis muscles may be seen in the lateral compartment, and the gastrocnemius muscle and soleus muscle may be seen in the posterior compartment.

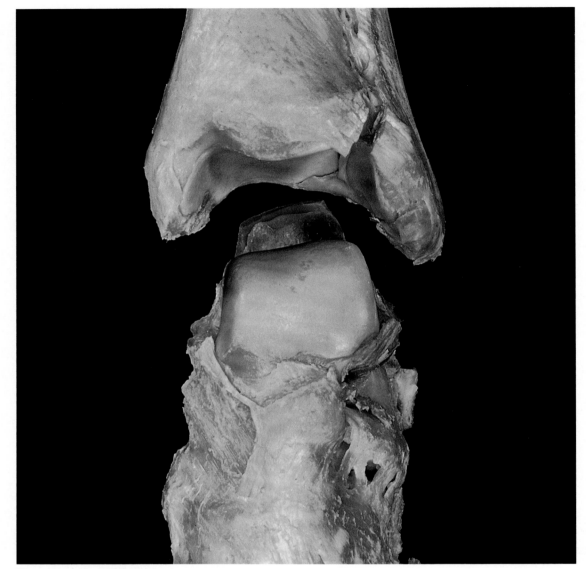

Figure 7.18A   Ankle joint.

The Lower Limb **99**

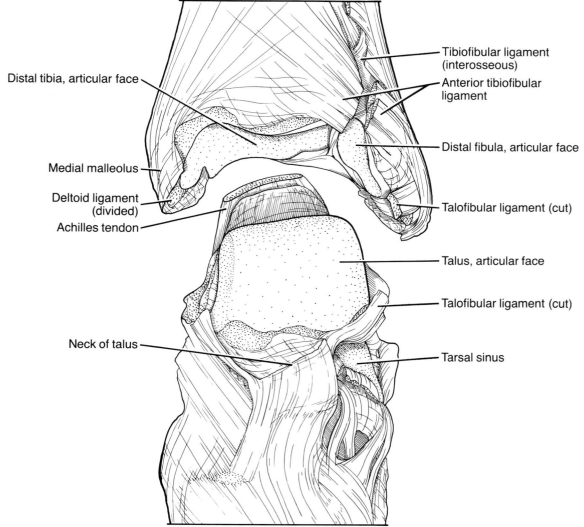

Distal tibia, articular face

Medial malleolus

Deltoid ligament (divided)

Achilles tendon

Neck of talus

Tibiofibular ligament (interosseous)

Anterior tibiofibular ligament

Distal fibula, articular face

Talofibular ligament (cut)

Talus, articular face

Talofibular ligament (cut)

Tarsal sinus

Figure 7.18B   Ankle joint.

Figure 7.18A, B   Ankle joint. An anterior view of the interior of the ankle joint shows the broad articular surface of the talus and its counterpart articular surfaces of the inferior tibia and the lateral malleolar portion of the fibula. Tibiofibular ligaments help to maintain the mortise into which the articular portion of the talus fits. The deltoid ligament medially and the talofibular ligament laterally are collateral ligaments of the ankle joint.

Systems Approach Grid

	Central Nervous System	Head and Neck	Thorax	Abdomen	Pelvis	Upper Limb	Lower Limb
Skeletal System		2.2, 2.3, 2.4	3.1		5.4, 5.9	6.5, 6.7, 6.8, 6.9, 6.11, 6.12, 6.13, 6.15, 6.16	7.1, 7.3, 7.10, 7.11, 7.12, 7.13, 7.14, 7.15, 7.18
Muscular System		2.1	3.1, 3.3	4.1, 4.2	5.1, 5.2, 5.8	6.1, 6.2, 6.3, 6.4, 6.5, 6.6, 6.7, 6.8, 6.9, 6.10, 6.11, 6.12, 6.13, 6.14, 6.15	7.1, 7.2, 7.4, 7.5, 7.6, 7.7, 7.8, 7.9, 7.10, 7.13, 7.14, 7.15, 7.16, 7.17
Nervous System	1.1, 1.2, 1.3, 1.4, 1.5, 1.6, 1.7, 1.8, 1.9, 1.10, 1.11	2.1, 2.3, 2.5	3.3, 3.4, 3.5, 3.6	4.2	5.1, 5.3	6.2, 6.6, 6.8, 6.10, 6.12	7.4, 7.5, 7.6, 7.7, 7.8, 7.9, 7.15, 7.16, 7.17
Endocrine System			3.2, 3.6	4.14, 4.15, 4.16	5.5, 5.6	6.8	
Cardio-vascular System	1.1, 1.2, 1.3, 1.6, 1.7, 1.8, 1.9, 1.11		3.2, 3.3, 3.4, 3.5, 3.6, 3.7, 3.8, 3.9, 3.10, 3.11	4.2, 4.3, 4.7, 4.10, 4.11, 4.14, 4.15, 4.16	5.2	6.6, 6.8, 6.10	7.3, 7.4, 7.5, 7.8, 7.15, 7.16, 7.17
Lymphatic System		2.1	3.5	4.6, 4.10, 4.11, 4.16		6.6	7.4, 7.8
Respiratory System		2.2, 2.3, 2.4	3.1, 3.2, 3.3, 3.4, 3.5, 3.6, 3.7, 3.8	4.1, 4.2, 4.3		6.3, 6.6	
Digestive System		2.2, 2.3, 2.4	3.3, 3.5	4.3, 4.4, 4.5, 4.6, 4.7, 4.8, 4.9, 4.10, 4.11, 4.12, 4.13	5.3, 5.4, 5.5, 5.6, 5.8		7.2
Urinary System				4.14, 4.15, 4.16	5.3, 5.4, 5.5, 5.6, 5.8		
Reproductive System			3.1		5.1, 5.2, 5.3, 5.4, 5.5, 5.6, 5.7, 5.8		

Index